APPLIED STATISTICAL SCIENCE

DEPENDABILITY ASSURANCE OF REAL-TIME EMBEDDED CONTROL SYSTEMS

APPLIED STATISTICAL SCIENCE

Additional books in this series can be found on Nova's website under the Series tab.

Additional E-books in this series can be found on Nova's website under the E-books tab.

APPLIED STATISTICAL SCIENCE

DEPENDABILITY ASSURANCE OF REAL-TIME EMBEDDED CONTROL SYSTEMS

FRANCESCO FLAMMINI

Nova Science Publishers, Inc.
New York

Copyright © 2010 by Nova Science Publishers, Inc.

All rights reserved. No part of this book may be reproduced, stored in a retrieval system or transmitted in any form or by any means: electronic, electrostatic, magnetic, tape, mechanical photocopying, recording or otherwise without the written permission of the Publisher.

For permission to use material from this book please contact us:
Telephone 631-231-7269; Fax 631-231-8175
Web Site: http://www.novapublishers.com

NOTICE TO THE READER

The Publisher has taken reasonable care in the preparation of this book, but makes no expressed or implied warranty of any kind and assumes no responsibility for any errors or omissions. No liability is assumed for incidental or consequential damages in connection with or arising out of information contained in this book. The Publisher shall not be liable for any special, consequential, or exemplary damages resulting, in whole or in part, from the readers' use of, or reliance upon, this material. Any parts of this book based on government reports are so indicated and copyright is claimed for those parts to the extent applicable to compilations of such works.

Independent verification should be sought for any data, advice or recommendations contained in this book. In addition, no responsibility is assumed by the publisher for any injury and/or damage to persons or property arising from any methods, products, instructions, ideas or otherwise contained in this publication.

This publication is designed to provide accurate and authoritative information with regard to the subject matter covered herein. It is sold with the clear understanding that the Publisher is not engaged in rendering legal or any other professional services. If legal or any other expert assistance is required, the services of a competent person should be sought. FROM A DECLARATION OF PARTICIPANTS JOINTLY ADOPTED BY A COMMITTEE OF THE AMERICAN BAR ASSOCIATION AND A COMMITTEE OF PUBLISHERS.

LIBRARY OF CONGRESS CATALOGING-IN-PUBLICATION DATA

Flammini, Francesco.
Dependability assurance of real-time embedded control systems / Francesco Flammini.
p. cm.
Includes index.
ISBN 978-1-61728-502-8 (softcover)
1. Embedded computer systems--Reliability. 2. Automatic control--Reliability. 3. Real-time control. I. Title.
TJ213.95.F53 2010
629.8'9--dc22
                              2010022715

Published by Nova Science Publishers, Inc. ✣ *New York*

CONTENTS

Chapter 1	Introduction: Critical Applications	1
Chapter 2	Critical Computer Systems	5
Chapter 3	Dependability Prediction Techniques	39
Glossary of Acronyms		51
References		55
Index		71

Chapter 1

INTRODUCTION: CRITICAL APPLICATIONS

Critical applications are defined as the ones whose failure has a relevant impact on:

- finance
- human beings
- environment

A failure occurs where the application is no more able to guarantee its required function. There exist several levels of criticality, according to the several possible classes of failure. The simplest distinction, for instance, is between the classes of failure corresponding to service unavailable or service incorrect, with many nuances within those two extremes (i.e. partial failures). Generally speaking, it is better to make the system unavailable when a possibly incorrect behavior is diagnosed, when system unavailability corresponds to a safer state. However, in other cases it is better to trust an even not completely reliable system, when no backup or fall-down procedure is available. Such a consideration implies that we have to reason in terms of failure consequences. A simple metric to express the criticality of a failure is given by the following expression:

$$EC_F = P_F * C_F \qquad (1)$$

Where:
EC_F is the expected cost of failure (sometimes defined as "Risk");
P_F is the probability of occurrence of failure;
C_F is the cost of failure.

Obviously, most critical failures correspond to higher C_F; however, the overall failure criticality is given by EC_F, which takes into account the frequency of occurrence of the failure. For instance, an application featuring just one failure mode with a totally negligible probability of occurrence cannot be defined as critical. This is the reason why going out when it is raining is not considered a critical application, as the probability of being stroked by a lightning is quite low (this is also the reason why nobody has invented a lightning discharging umbrella yet...). This also gives an idea of where to intervene to improve the trustworthiness of an application.

While it is quite straightforward to measure financial losses due to service unavailability or malfunction, it is quite hard to reason in terms of environmental impact or even human losses. For the latter aspect, someone tried to achieve such an objective by introducing the ALARP (*As Low As Reasonably Practicable*), MEM (*Minimal Endogen Mortality*) and other apparently odd criteria for risk management (assessment and consequent decision making).

There exist a number of critical applications. A rough division can be performed between safety-critical (or life-critical, as the so called catastrophic failures can have an impact on human beings) and mission-critical (or money-critical, as failures have only a financial impact, neglecting possible consequent suicides...) ones:

- Defence, aerospace, biomedical apparels, transportation, industrial processes, chemical and nuclear plant control are all examples of safety-critical applications;

- Web servers, network routers, internet backbones, transaction modules, bank information systems, satellites, communication facilities, broadcasting services are instead examples of mission-critical applications.

Usually, a safety-critical application is mission-critical as well (see Figure 1). This is the reason why we concentrate on safety-critical applications and related control systems, as they represent the most complete case-studies.

Since now, we have not referred specifically to computer-based control systems, as critical applications are not necessarily controlled by computers. However, as application complexity grows up, computers reveal much more reliable than human beings in performing control actions. This is the reason why a great part of the computer research community is devoted to this field. Despite of the great variety of proposed and implemented techniques, catastrophic failures continue to occur (see e.g. [1]).

Chapter 2

CRITICAL COMPUTER SYSTEMS

Critical computer systems are meant to control critical applications [2]. Such systems can be safety-critical and/or mission-critical, the latter class being usually included in the former. The general scheme of a control system is provided in Figure 2; in a computer-based control system, the control logic is implemented using software programs and hardware circuits. A failure of the computer-based control system is very likely to have catastrophic consequences; however, this is not always true, for the following two reasons:

1. Even safety-critical systems feature non critical failure modes; this is why it is important to analyze and classify failures and how they propagate by means of proper *Failure Mode Effects and Criticality Analysis* (FMECA) [3];
2. Simpler fall-back systems or manual procedures can still be available to control the application.

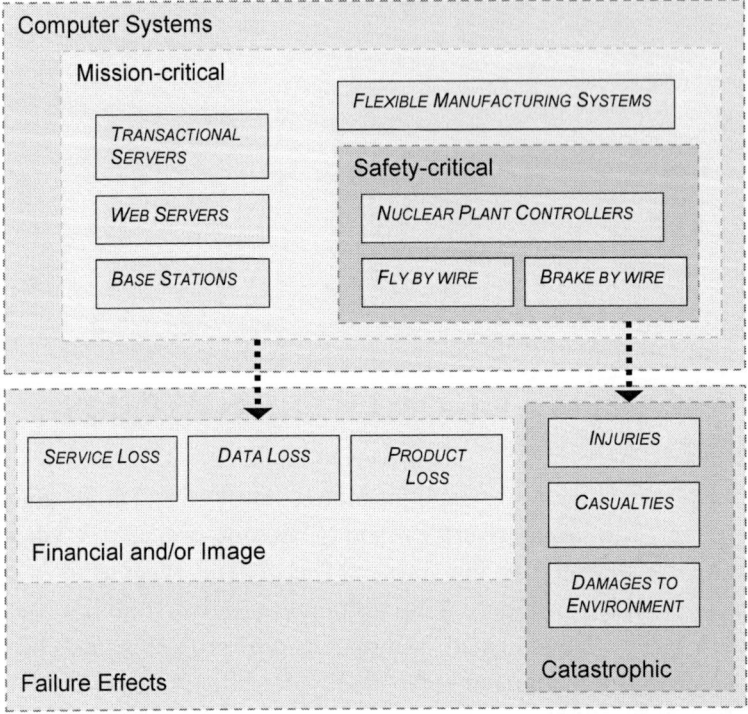

Figure 1. Classification of critical computer systems with respect to failure effects.

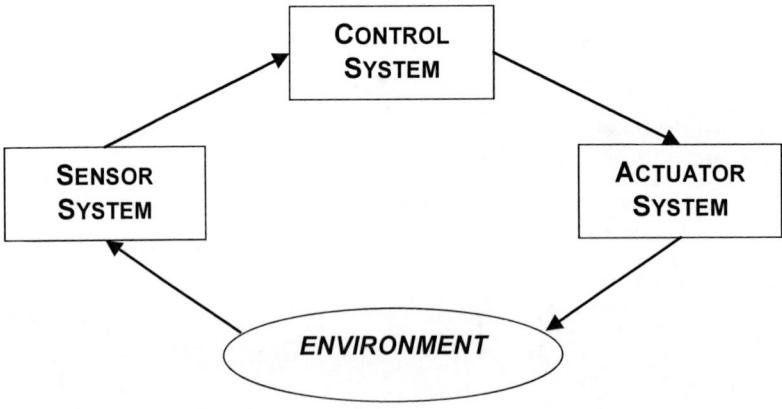

Figure 2. General scheme of a control system.

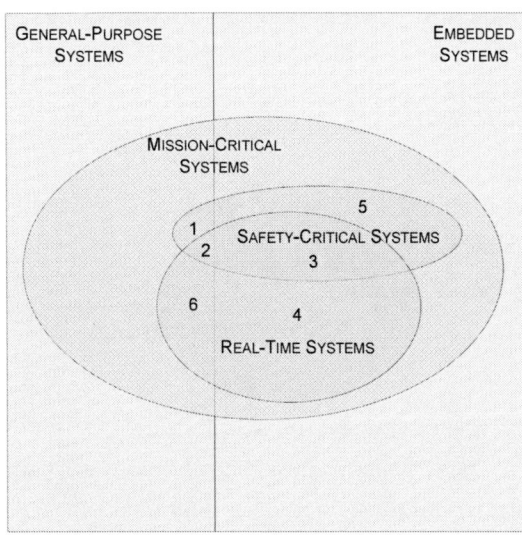

Figure 3. A classification of critical computer systems with respect to their architecture.

Computer based control systems are usually designed to be embedded (see Figure 3), that is dedicated to the specific application, as this allows to ensure that their architecture is as simpler as possible for higher survivability and easy verifiability, according to the most elementary *design for testability* criteria. As control systems have to operate in real-time, the related computing systems have to provide mechanisms to allow for time predictability. For instance, if the minimal time constant of the application is τ, the computing cycle T_C must be significantly lower then τ, assuming all control processes are scheduled and complete their operations within a single computing cycle. A failure in respecting real-time constraints is defined as a *timing failure*. A timing failure can be determined by performance factors, design flaws (e.g. deadlocks) and misuse (e.g. overloads). As for any computing system, service alterations (i.e. failures) can be determined by a number of causes (i.e. faults) injecting state alterations (i.e. errors) which can propagate until reaching system interfaces, after a possible varying latency. A classification of faults, errors and failures is provided in the general definition of *dependability* recently provided by Laprie and others [4]. An informal and synthetic definition of dependability is the following: "Dependability is the ability of a system to provide a service that can justifiably be trusted". A more formal definition of dependability integrates the following main concepts:

- ATTRIBUTES: Availability, Reliability, Safety, Confidentiality, Integrity, Maintainability
- THREATS: Faults, Errors, Failures
- MEANS: Prevention, Tolerance, Removal, Forecasting

Safety and Availability are by far the most important attributes for critical control systems. A safe system must behave correctly in any operating and environmental conditions, including degraded ones. At the service level, Safety is therefore a functional aspect, but safety assurance requires the implementation of structural mechanisms both at hardware and software levels in order to detect and tolerate potentially hazardous faults. Such mechanisms are known as *fault tolerance* techniques and are included in the MEANS of the general concept of dependability. For certain classes of systems, Safety and Availability are strictly correlated, as an unavailable system can impact on application safety, as it happens in aerospace. Availability is always strictly correlated to Reliability and Maintainability, as the first expresses the continuity of the service, the latter the capacity of restoring it after it has been interrupted: together they give a measure of how long the system is available for use (i.e. providing a correct service), in absolute or relative terms. For instance, reasoning in terms of mean values, the steady state availability A of a system being in only two states (Available / Unavailable) is given by the following expression[1]:

$$A = \frac{MTBF}{MTBF + MTTR} \qquad (2)$$

---

[1] The formula is obtained by solving the two states Markov Chain as explained in [5] (see also Figure 11).

Where:

MTBF is the Mean Time Between Failures
MTTR is the Mean Time To Repair

In this work we concentrate on the MEANS of Failure Forecasting, which can be performed using both simulative and formal techniques, starting from fault models (type and probability of occurrence). The context diagram of Figure 4 synthesizes the reference context in which a critical computer-based control system operates and the related threats to its dependability in terms of fault sources (all interacting entities are potential causes of malfunctions). For instance, system engineers are likely to inject requirements or coding faults leading to latent errors and/or systematic failures, while users can inject both interaction and malicious faults (i.e. hackers' intrusion). In particular, for safety-critical systems the causes of potentially dangerous malfunctions are analyzed, classified and mitigated during *hazard-analysis* sessions (usually performed by heterogeneous groups of experts using informal brainstorming-based approaches).

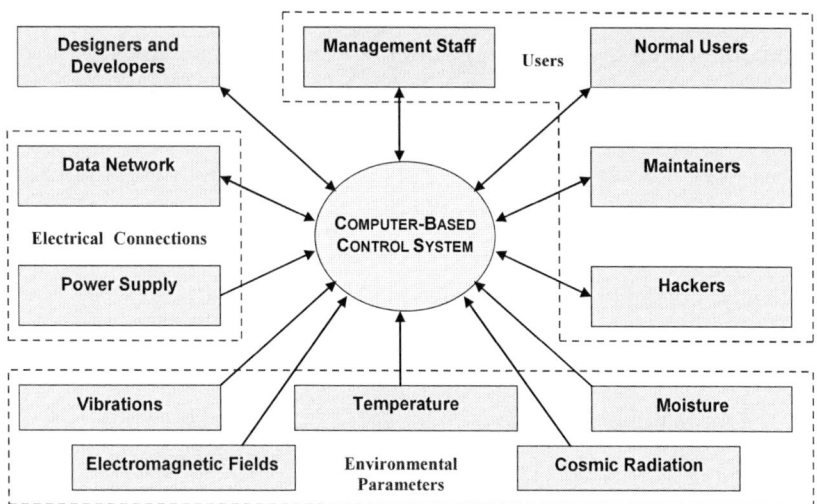

Figure 4. Reference context diagram showing external dependability threats.

2.1. REFERENCE ARCHITECTURES

Two main classes of computers exist: general purpose and embedded ones. General purpose systems are meant to be used in a variety of usually non critical applications, featuring standard hardware and operating systems. However, they are not suited to be used as industrial controllers for the following reasons:

- Their hardware and software architecture is hardly analyzable and thus the related behavior is not predictable;
- They are developed with the main aims of low cost and high performance, which usually imply low reliability and survivability;
- They do not implement any fault-tolerance mechanism at the hardware level, therefore safety must be assured at the application software level, causing usually unacceptable efforts and/or overheads.

Despite of such considerations, even commercial general purpose computers can be used in critical applications; when this happens, software redundancy and fault-tolerance mechanisms are stressed in order to ensure system dependability. Software fault tolerance mechanisms, usually based on temporal redundancy (in opposition to spatial redundancy, which is usually implemented in hardware), include: recovery blocks, n-version programming, etc. [6].

However, specific hardware and software configurations (i.e. embedded architectures) are by far the most widespread in real-time systems design. They are often classified basing on the CPU level redundancy, i.e. majority voting mechanism; we can find:

- *2 out of 2 (2oo2) architectures;* they constitute the simplest way to significantly lower the probability of undetected errors due to CPU computations and/or memory access; however, system reliability also decreases substantially, as the two CPU constitute a series system (system is no more available as soon as the first CPU fails);

Critical Computer Systems

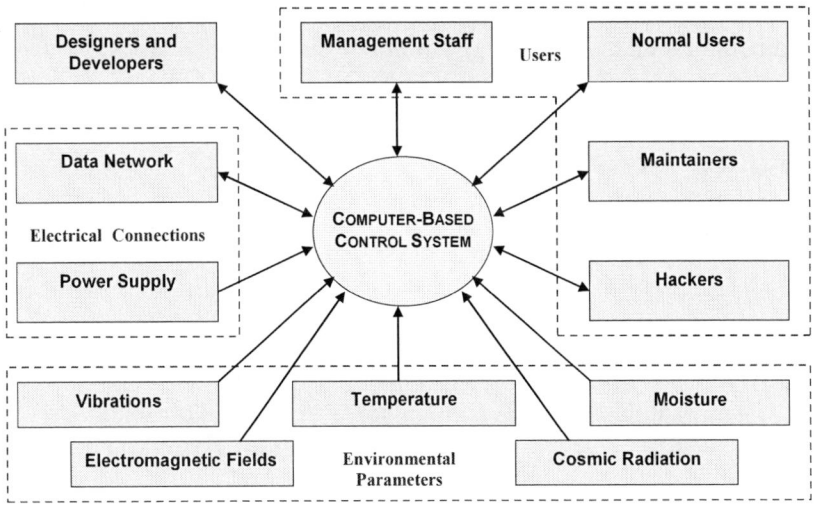

- *2 out of 3 (2oo3) architectures,* also known to be based on a widespread Triple Modular Redundancy (TMR) scheme, increase by far system availability, as they can continue working in a 2oo2 fall-back configuration when a single CPU fails. While intuition suggests that the probability of 2 CPU providing the same erroneous result increases in these architectures with respect to 2oo2, the expense in terms of safety is negligible;
- *K out of N architectures* (with K<N) constitute quite uncommon hardware configurations sometimes used in aerospace applications, e.g. space shuttles; one advantage is that the K/N ratio can be fine tuned to meet specific safety and availability requirements, the latter being essential for mission-critical applications; another advantage is that commercial inexpensive and less reliable machines can be used, while leaving unvaried system dependability attributes, at a major expense in terms of voter complexity and criticality.

Typically, a so called "watchdog" monitors section outputs and processes heartbeat signals, shutdowning sections whenever a fault is detected (which can also be a timing fault, in case a software stall due to a deadlock happens).

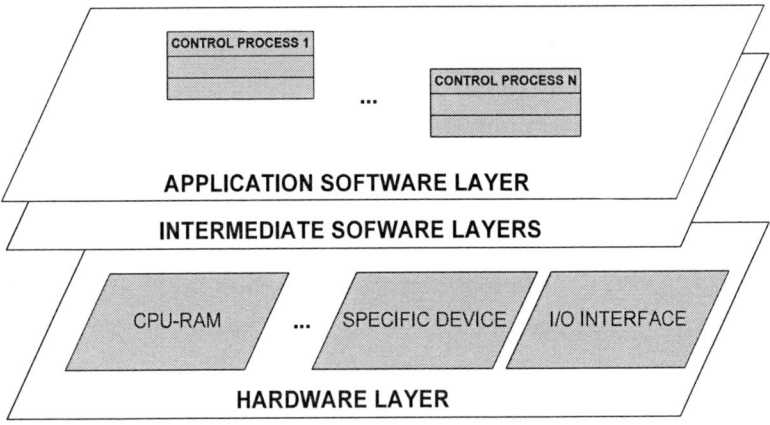

Figure 5. Abstraction layers of a control system's constituent.

Despite of the specific architecture used for the CPU-memory subsystem, real-time embedded computers are based on a "no-single-point-of-failure" reliability engineering approach, that is all critical components and subsystem are redundant. Less reliable components, like power supplies, can be replicated several times, in a hot stand-by configuration (i.e. backup units become active without service interruption). Central memories, besides being replicated, usually feature Error Correcting Codes (ECC) circuits for automatic error detection and correction on one or more memory cells. Mass-storage subsystems, whenever required, are based on RAID disk arrays, with spare units in hot stand-by configuration. For all necessary components the main choice is about "make or buy": they can be developed ex novo in order to meet specific reliability and robustness requirements or can be selected in the COTS (Commercial Off The Shelf) market. Finally, maintainability strategies are very important to minimize down-time and thus ensure high-availability, including scheduled preventive maintenance (to avoid components' wear-out and restore possibly degraded redundancy), quick diagnosis, repair and recovery actions and proper management of spare components; system architecture must also accommodate for efficient maintenance.

The characteristic of being *distributed* is common in industrial control systems, as sensors and actuators are often needed to be installed far from the logic controller, also featuring autonomous functionalities (e.g. self diagnostic and repair facilities). Moreover, many intelligent transportation systems (e.g. fly-by-wire, brake-by-wire, railway interlocking, etc.) are needed to be geographically distributed, featuring a fixed (usually the "ground" part) and a moving part, e.g. installed on planes', ships' or trains' cockpits. Therefore, most control systems feature many diverse and distributed devices.

A more detailed scheme of a single subsystem is provided in Figure 5. The underlying hardware layer is usually implemented connecting device cards by means of a proper backbone (e.g. a VME industrial bus [7]). A single card, e.g. CPU-RAM, is connected into a slot to be easily hot-swappable; it is thus called a Line Replaceable Unit (LRU). Embedded systems' hardware includes: Power Supplies (PS) units; Direct Memory

Access (DMA) controllers; PLC (Programmable Logic Controllers) chips; Digital to Analogical and Analogical to Digital Converters (DAC/ADC), needed for the system to interact with the external environment (which is analogical in its nature); several other specific devices, interface cards and communication buses. Intermediate software layers represent Operating System tasks (e.g. CPU scheduler, system diagnostic processes, etc.) and layers of the ISO/OSI protocol stack (transport, networks, data link, etc.), providing the necessary inter-system communication facilities in the distributed environment. There exist a number of commercial already available and certificated real-time operating systems for embedded computers, e.g. VxWorks [8]. The application software layer can also be divided into more layers, e.g. when a dedicated underlying layer is found to be appropriate in order to better manage system operating modes. Finally, at the application layer, variously implemented control processes realize the core logic of the subsystem.

Critical systems are designed to be decomposed in *fault confinement regions*, that is architectural or functional modules such that their Error Detection and Recovery Mechanisms (EDRM) avoid that an internal error propagate to other modules, thus satisfying the fault isolation property of fault-tolerance.

In [55] a very interesting research work is presented dealing with a generic fault-tolerant architecture for real-time dependable systems (GUARDS project).

2.2. LIFE-CYCLE MODEL

The life-cycle model of real-time systems is not different from the one of any other hardware and software system. The novelty for embedded systems is the possibility of model-based hardware and software *co-design*, supported by specific tools (e.g. the Rational Rose Real-Time software suite, supporting OMG Model Driven Architecture paradigm [9]). In fact, differently from commercial general-purpose systems, where design only regards the application software layers, embedded systems hardware architecture has often to be designed from scratch, in order to meet specific requirements (size, shape, survivability, etc.). A possible *waterfall development model* (which is more realistically a spiral one in practice) is reported in the following:

- Phase 0: Feasibility Study
- Phase 1: Analysis of Requirements
- Phase 2: System Design
- Phase 3: System Implementation
- Phase 4: System Testing
- Phase 5: Operational Life
- Phase 6: Dismission

A slightly different and variously customized version of the above waterfall model can be found in any safety standard. Phase 0 (regarding above all the choice to make a new system, buy or adapt an already available and suited one or continue using the existing one) and phase 6 (in which system is heavily upgraded or substituted for obsolescence or other reasons) are not of great interest for this work. Phases 1 to 5 are instead of great interest and are almost identically critical in real-time system development. The development cycle is represented by a spiral model in practice, as each phase is retrofitted with the result of the following ones and the overall process is iterative. Reference [10] describes a specific model for critical software development which allows ensuring quality and safety integrity levels.

2.2.1. Analysis of Requirements

The definition of requirements is usually performed by the supplier together with the customer. Functional (FRS, Functional Requirements Specification) and non functional (e.g. RAMS, ergonomics, performance, etc.) requirements are agreed by analyzing first of all basic needs of the customer, and then more particular and detailed operating scenarios, strictly related to the specific application. Many requirements refer to existing national or international reliability and safety standards, which can be mandatory for any new safety-critical system developed in that country. Laws and norms play an important role in this phase. Functional safety-requirements are defined by a preliminary hazard-analysis, but they are added during the entire life cycle of the project as more detailed hazard-analysis sessions (e.g. integration or subsystem HA), are possible. System level requirements are usually expressed informally, using natural language syntax.

2.2.2. System Design

The system has to be properly designed before any implementation is started. This involves an architectural design which defines and describes the main blocks and components of the system, their interfaces and interactions. The needed hardware is defined and the software is split up in its components. Software components have to be defined to meet end user requirements as well as possible scalability needs. The aim of this phase is to generate a System Architecture Document which serves as an input for software and hardware design or selection. In practice, system engineering requires the definition of the hierarchy of documents from SRS (System Requirements Specification), in which high level system architecture is defined, to:

- SSRS (Sub-System Requirements Specification), in which system requirements are allocated on the single subsystems of the possibly distributed architecture;
- FIS (Functional Interface Specification), in which interface requirements are defined to allow for the interoperability of subsystems, in terms of correct interpretation and management of exchanged information (data representation and semantic, communication protocol, etc.);

- HWRS (Hardware Requirements Specification), in which requirements and specific functions are allocated to the hardware layer;
- HDS (Hardware Design Specification), containing the hardware architecture description of the various subsystems; module functions, characteristics and configuration are defined by block schemes;
- SWRS (Software Requirements Specification), in which requirements and specific functions are allocated to the software layer;
- SAS (Software Architecture Specification), regarding the distribution of functionalities among software modules;
- SDD (Software Design Description), providing the detailed low-level documentation of software modules.

The ones listed above are just examples of the specification hierarchy which usually includes much more design documents. All requirements in the various documents are usually classified according to a set of keywords (e.g. functional/non functional, hardware/software, testable/non testable, etc.) and traced between documents (specific traceability analyses are performed). Safety requirements, being critical, are produced by means of specific hazard-analysis sessions, in which a more systematic approach (e.g. HazOp, see [11]) is used in order to detect and mitigate possible causes of incidents. Requirements can be specified by translating the high level natural language specification into formal languages like B and Z (see [12]). The use of formal means of specification, implementation and analysis during the entire development cycle of the system is highly recommended by international standards, as they provide sound means to detect and eliminate specification ambiguity, incoherence and incompleteness. Another advantage is the possible automation of the development process, provided that validated tools exist which can perform such an operation. However, entirely specifying a complex system by means of formal languages is unfeasible in practice; moreover, even whether feasible, no available tool would be so efficient to check such specification. Therefore, in industrial practice only small subsystems or protocols are specified and verified by means of formal methods (see for instance [13]).

The management (maintenance, numbering, classification, traceability, etc.) of requirements is also a critical task which is usually supported by specific requirements management tools, like Rational Requisite Pro [14] or Telelogic Doors [15].

System level requirements are usually provided by company "engineering" units, while lower level requirements are specified by experts of "development" units, who also entirely manage the following phase.

2.2.3. System Implementation

Development, which means coding for software and production for hardware, requires respecting quality standards, which are quite severe for critical systems. For instance, when a general purpose programming language is employed, rules are defined in order to avoid possibly ambiguous and then dangerous syntactical constructs or statements (only a subset of its syntax is therefore employed). In general, software architecture must be kept as linear as possible, in order to easily control its structure and behavior, facilitate static analysis, code walk through, dynamic verification, diagnosis and maintainability. Therefore, well-structured and object-oriented software design and programming paradigms are highly recommended. This sometimes implies a negative impact on performance, which is usually negligible for real-time systems as far as timing requirements are still demonstrated to be met. Another way to express the same concept is that controllability and predictability are much more important than performance for real-time systems. This is the reason why whenever possible processes' scheduling is chosen to be fixed, with no possibly risky preemption (which could lead to deadlocks). *Defensive programming* checks are also recommended in order to avoid overflows or other errors leading to unpredictable behaviors when control variables are assigned values out of the range of their specified domains. Compilers and other development tools have to be validated as well in order to trust their output (this is the reason why automated model-based development tools like Rational Rose Development Studio [91] based on a proprietary RT UML), which are widespread for commercial embedded systems development, are rarely employed for critical systems' development). All these techniques belong to the class of fault prevention.

As for software fault tolerance, many techniques exist, some of them have already be cited above; the redundancy is usually based on a *diversity* approach: two or more diverse implementation of the same module, produced by independent development teams using different tools and running sequentially (i.e. in temporal redundancy configuration) or in parallel (i.e. in a spatial redundancy configuration), have to agree on results for them to be accepted (a hardware-like majority voting is also possible). The choice of what mechanisms to stress and at what level to implement them (hardware and/or software) is usually fine tuned considering the needs and characteristics of the specific application (this is one aspect of co-design).

As an example, for railway systems it is usually possible to reach a safe state corresponding to train standstill, therefore software fault-tolerance is not stressed as far as an underlying safe hardware platform (e.g. a TMR) is available; in fact, software errors are assumed not to be generated by hardware malfunctions (when no hardware faults are detected by scheduled diagnostic processes) and therefore can be simply managed by reaching the safe state (e.g. system or section[2] shutdown) after detection, without any roll-back or other software recovery action. Of course, this is not possible when an untrustworthy hardware platform is available (e.g. a mono-processor commercial machine); in such a case software fault tolerance is necessary.

As for hardware, traditional fault prevention (de-rating, use of military highly survivable components, burn-out tests, etc.) and tolerance (use of

---

[2] A "section" is defined as a hardware module whose output is to be voted together with the one of other sections. It is generally constituted by a CPU card, possibly comprising local memories and other devices.

redundant back-up components in cold or hot stand-by, in passive or active configurations, e.g. voting) mechanisms are implemented. In particular, it is important to electrically decouple components (e.g. using fiber-optics cables for components' interconnection and different power supplies for independent sections) to provide fault containment regions, in order to reduce common mode failures. Fault removal has to be ensured by means of hardware checking diagnostic processes which have to be scheduled regularly by the operating system, together with application processes; moreover, self diagnosis and alert facilities have to be provided as well as hot swapping mechanisms, at least for less reliable components, in order to reduce the mean down time (time to diagnose, repair or substitute, and restart). The effectiveness of all such mechanisms is also checked in the following testing phase.

2.2.4. System Testing

Intuitively, testing plays an essential role in critical systems' development. With respect to other types of systems, for real-time systems we can distinguish two levels of testing:

- Engineering/development tests, aimed at ensuring that the system works as predicted; the objective here is just a rough check that design choices are correct and well implemented;
- Verification & validation tests, aimed at ensuring an extensive coverage of system functionalities in order to demonstrate that the product can be certified to be compliant to international safety and reliability standards.

V&V also includes the check of requirements for consistency (unambiguity, completeness, coherence) and correctness, using both informal and, whenever feasible, formal means of analysis.

Hardware verification is usually performed by:

- Stress tests, aimed at checking the robustness and survivability of components to adverse environmental conditions;
- Fault-injection (either physical or simulated, by means of HDL models), aimed at checking the coverage of fault-tolerance mechanisms as well as providing statistical measures of components' reliability by means of accelerated tests;
- EMC (Electro Magnetic Compatibility) tests, aimed at verifying components' rejection to electromagnetic fields (in other words, their screening against radiation is checked).

Formal methods can be used to statically evaluate the probability of occurrence of hazardous events (i.e. THR, total hazardous rate, or MTBHE, Mean Time Between Hazardous Events) or to estimate system availability, e.g. by means of Fault Tree Analysis, Markov Chains models or Stochastic Petri Nets (see [134] for a set of interesting applications).

Software Tests Comprise:

- Module testing, aimed at verifying the single software module (e.g. a specific function);
- Integration testing, aimed at verifying that modules interact correctly and cooperate, running on the underlying hardware, in order to implement the desired macro-functions;
- System testing, in which hardware and software of the system under test are verified together, as a whole, using a black-box approach; in other words, the overall system is verified against its functional requirements (this is the reason why this process is also known as functional testing[3]).

Besides such tests, usually targeted to the application software, more specific logic and configuration tests are performed using both static and

---

[3] As code-generated software errors are permanent, functional testing belongs to the class of systematic defect testing; hardware-generated software errors are instead known as soft errors (e.g. single bit errors in memory cells).

dynamic approaches (the latter requiring code execution). Static approaches do not require code execution and can be performed by hand, using automatic tools or even formal means of analysis (e.g. model checking or theorem proving; see [122]). Dynamic approaches are instead based on simulation. Finally, acceptance and on-the-field tests are performed on the real system installation, and are usually based on a limited and feasible subset of system test-suite (which is entirely executed in laboratories, where also abnormal conditions can be easily implemented).

2.2.5. System Operational Life

The useful life of the system begins after it is put in exercise and begins to be operational. Critical actions in this sense regard correct installation, on-the-field tests, pre-exercise, and monitoring. It is a matter of fact that regardless of a thorough laboratory-based verification phase, unpredicted problems arise when the system is installed in the real environment. From the functional point of view, systematic failures often depend on incorrect configuration, which is not coherent with on the field data; a rate of random faults which is higher than predicted can be instead a consequence of an imprecise estimation of threats; finally, ergonomy can be detected to be inadequate when typical users interact with the system, so interfaces have to be retouched. The management of repair actions and of spares is also critical because strictly related to logistics and human factors. Moreover, even after the pre-exercise phase is successfully passed, a statistical monitoring of critical reliability parameters is needed; such process is known as FRACAS (Failure Reporting Analysis and Corrective Action System) [124], and also comprises the definition of corrective actions needed to face the detected anomalies, besides a rigorous and systematic monitoring, classification and analysis of operational faults. Besides corrective maintenance actions, also adaptive and improvement maintenance is usually required during the operational life of the system in order to cope with specific needs, which were not predicted during system design. It is important to underline that after a maintenance action of any type (e.g. leading to a new improved software version) non regression testing is needed, aimed at verifying that modifications have not impacted negatively on other functionalities, even though the latter were already verified and found to be perfectly working. Another often underestimated threat comes from reuse of library routines which did not undergo extensive system testing, as in the case of the famous Ariane 5 satellite launcher failure (see [1] for an interesting survey of catastrophic failures).

After many years of successful operation of different installations with no detection of systematic failures, the system together with its components and its development environment are said to be "proven in use" (which is an informal but highly perceived demonstration of product trustworthiness).

For critical systems, the correct management of actions in each development phase as well as the passage from one phase to another is strictly regulated by norms provided by international verification and validation standards, which are the object of next section.

2.3. VERIFICATION AND VALIDATION STANDARDS

Safety-critical industrial control systems need to be certified by an independent assessor before they can be put into service. The role of the assessor (e.g. TÜV, see [149]) is to check that procedures are performed in compliance to safety standards; after a number of safety audits, possibly involving customer representatives, the product is declared to be in conformity with the applicable standards. The certification is performed according to different standards and norms, depending on the nature of the system and as required by (inter)national laws. Standards provide general guidelines as well as mandatory requirements about rules to be respected, activities to be performed and documentation to be produced during the entire life-cycle of the project. For instance, with respect to critical software development and verification techniques, we can find the following attributes (refer to [CENELEC 50128]:

- Mandatory (M), e.g. functional testing;
- Highly Recommended (HR), e.g. use of formal engineering methods;
- Recommended (R), e.g. diversity;
- Not Recommended (NR), e.g. neural networks (for unpredictability reasons).

Such attributes are dependant from the Safety Integrity Level (SIL) required by the specific subsystem. In fact, a complex critical system is in general constituted by both critical and non critical apparels. A non critical apparel can be, for instance, a visual management subsystem whose unavailability only impacts on performance. The SIL level for such a subsystem can be chosen to be lower. Usually, SIL levels go from 0 to 4. SIL 0 apparels are not safety critical in any way and can be developed using the same approaches of commercial products. On the opposite extreme, SIL 4 represent the most sever level of safety integrity, which is required for all subsystems whose output has a direct impact on system safety. The choice of the SIL level to be used can be a result of a preliminary hazard-analysis or simply a customer's constraint.

As mentioned above, many standard and guidelines exist for developing critical systems, and they are usually application specific, besides being distinguishable into hardware and software specific. In Table 1we report a not exhaustive list of well-known standards, together with a brief description of them, which are presently adopted in a variety of applications.

The required safety level for all standards is determined from the safety assessment process and hazard analysis by examining the effects of a failure condition in the system. For instance, in DO-178B [137], the failure conditions are categorized by their effects on the aircraft, crew, and passengers:

Table 1. Some standards and guidelines for reliability and safety critical systems

Name	Producer	Field	HW/SW	Description
IEC 61508	International Electrotechnical Commission	Aerospace	SW	Software safety in aerospace applications
DO-178B	Radio Technical Commission for Aeronautics (RTCA)	Aeronautics (Civil Avionics)	SW (DO-254 exist for HW)	Software Considerations in Airborne Systems and Equipment Certification
CENELEC 501XX	European Committee for Electrotechnical Standardization	Railways	HW+SW	Railways safety and reliability standard
YELLOW BOOK	Rail Safety and Standards Board	Railways	HW+SW	Engineering safety management in new railways' technologies
MIL-HDBK	U.S. Department of Defense	Any	HW	Electronic reliability design standards
IEEE V&V Standard 1012	Institute of Electrical and Electronic Engineers	Any	SW	Software verification and validation standard

- Catastrophic - Failure may cause a crash.
- Hazardous - Failure has a large negative impact on safety or performance, or reduces the ability of the crew to operate the plane due to physical distress or a higher workload, or causes serious or fatal injuries among the passengers.
- Major - Failure is significant, but has a lesser impact that a Hazardous failure (for example, leads to passenger discomfort rather than injuries).

- Minor - Failure is noticeable, but has a lesser impact than a Major failure (for example, causing passenger inconvenience or a routine flight plan change).
- No Effect - Failure has no impact on safety, aircraft operation, or crew workload.

All standards provide guidelines and requirements for processes to be performed and documents to be produced in the following areas:

- Planning
- Development
- Verification
- Configuration management
- Quality Assurance
- Certification

In the following sections we will provide general considerations about the definition of safety requirements, the assignment of criticality indices, the development model used for certification purposes, the reference V&V and RAM indices, and then we will present a specific railway-based standard.

2.3.1. Safety Requirements

Safety standards are very similar in most parts, but being each one developed independently from each other, often the terminology can be different. For instance, IEC 61508 [92] distinguishes two types of safety requirements: safety function requirements and safety integrity requirements. The safety function requirements govern the input/output sequences that perform the safety-critical operation. For example, a boiler could have a pressure sensor (input) that can reach a maximum value before the gas is shut off (output) to the burner. The safety integrity requirements of a system are composed of diagnostics and other fail-safe mechanisms used to ensure that failures of the system are detected and that the system goes to a safe state if it's unable to perform a safety function. Examples of integrity elements in the boiler would be a current range diagnostic on the pressure sensor or a watchdog timer. If either of these elements detected a failure they would be able to force the system to a safe state.

In general, all standards recognize that not every element of a system has the same effect on safe operation and therefore they allow some modules to be justified as independent. For this reason, the developer must employ a modular design method and define clear interfaces and protection mechanisms between those modules, in order to definitively classify subsystems into critical and non critical categories. This can be performed by using hardware memory protection through a memory-management unit or by using a language that enforces such encapsulation (Ada, Java, Modula, and so forth); the strength of the protection required depends on the SIL. The SIL of the entire system is then determined by constituents' SIL.

2.3.2. Criticality Analysis

One method for determining the SIL from the interaction among components is called a criticality analysis. The method for this analysis can be used for software components and the system as a whole. Each module is classified into one of four criticality levels:

- C3, Safety Critical: a module where a single deviation from the specification may cause the system to fail dangerously;
- C2, Safety Related: a module where a single deviation from the specification cannot cause the system to fail dangerously, but in combination with the failure of a second module could cause a dangerous fault;
- C1, Interference Free: a module that is not safety critical or safety related, but has interfaces with such modules;

- C0, Not Safety Related: a module that has no interfaces to safety-related or safety-critical modules.

At the simplest level, any module that's directly used in implementing a safety requirement would be C3, and any safety integrity requirements would be C2. Further classification of criticality and SIL design constraints would have to come from a detailed analysis of the design.

2.3.3. V-model

Standards for safety-critical software development suggest following a *V-model* development process. The evolutionary V-model shown in Figure 6 depicts the necessary connection between requirements and validation throughout the entire development process.

Some standard (e.g. IEC 61508) recognize that the V-model is not linear as several iterations of design and implementation may be necessary while end users and developers refine the requirements. Finally, software verification in the V-model could be further decomposed into the three sequential sub-steps: module testing → integration testing → functional testing (incremental verification).

With respect to the V-model, a more accurate distinction could be stated between dependability prediction and evaluation: dependability prediction is related to the "downward oriented activities" (i.e. design), while dependability evaluation is related to the "upward oriented activities" (i.e. verification). Despite of such distinction, there are no significant methodological differences between them: the main difference stands in the necessity for the former technique to evaluate more design options, needing for more flexible (e.g. parameterized) models, and to use only estimated data for a set of parameters (e.g. performance indices, which are not measurable as system is not yet implemented).

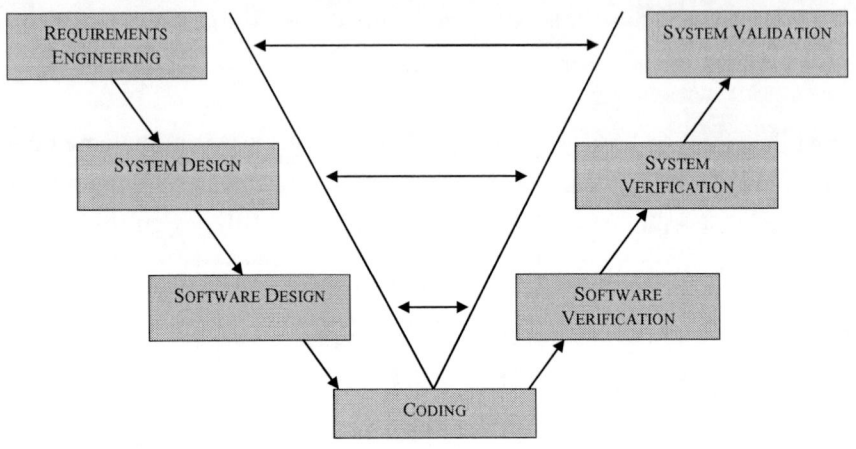

Figure 6. V-development model.

2.3.4. Verification and Validation Measures

Anomaly density measures can provide insightful information on the software product quality, including software development processes, and the quality of the V&V effort to discover and correct anomalies in the system/software. Anomaly measures and trends can be used to improve the quality of the project, and can be used to improve the planning and execution of V&V processes for future projects with similar characteristics. The measures are defined for the four development phases: Specification, Design, Implementation, Test.

The management of V&V activity uses measures to provide feedback for the continuous improvement of the V&V process and to evaluate the software development processes and products. Trends can be identified and addressed by computing evaluation measures over a period of time. Threshold values of measures should be established, and trends should be evaluated to serve as indicators as to whether a process, product, or V&V task has been satisfactorily accomplished. No standard set of measures is applicable for all projects so the use of measures may vary according to the application domain and software development environment. Even though no consensus exists on measures for evaluating the quality and coverage of

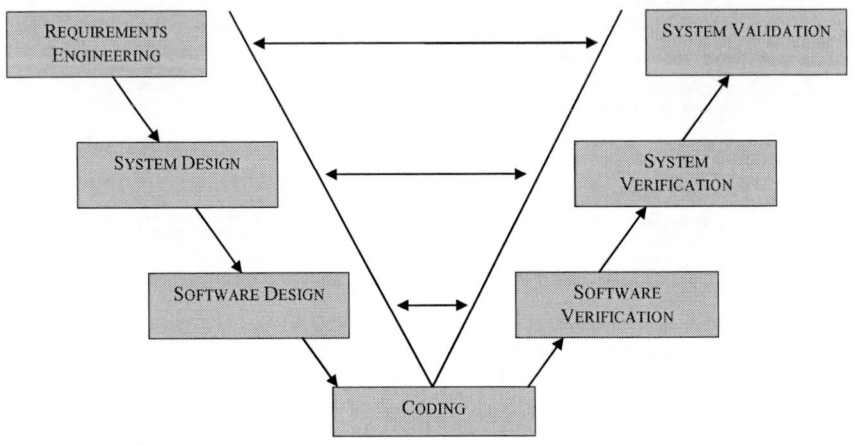

the V&V tasks, it is interesting to cite here three categories of measures which can be associated with the V&V effort (see [33]):

(1) Measures for evaluating anomaly density,

$$\text{X Anomaly Density} = \frac{\text{\# X Anomalies found by V \& V effort}}{\text{\# X reviewed by V \& V effort}}, \text{ where } X = \begin{cases} \text{Requirements} \\ \text{Design} \\ \text{Implementation} \\ \text{Test} \end{cases}$$

(3)

(2) Measures for assessing V&V effectiveness,

$$\text{X V \& V Effectiveness} = \frac{\text{\# X Anomalies found by V \& V effort}}{\text{\# X Anomalies found by all sources}}, \text{ where } X = \begin{cases} \text{Specification} \\ \text{Design} \\ \text{Implementation} \\ \text{Test} \end{cases}$$

(4)

(3) Measures for evaluating V&V efficiency,

$$\text{X V \& V Efficiency} = \frac{\text{\# X Anomalies found by V \& V in X Activity}}{\text{\# X Anomalies found by V \& V in all Activities}}, \text{ where } X = \begin{cases} \text{Specification} \\ \text{Design} \\ \text{Implementation} \\ \text{Test} \end{cases}$$

(5)

2.3.5. A Railway Based Example: CENELEC Standards

CENELEC is the European Committee for Electrotechnical Standardization [50]. CENELEC produced the following standards which are applicable to safety-critical railway control systems:

- EN 50126: Railway applications - The specification and demonstration of Reliability, Availability, Maintainability and Safety (RAMS);
- EN 50128: Railway Applications - Communication, signalling and processing systems - Software for railway control and protection systems;

- EN 50129: Railway applications - Communication, signalling and processing systems - Safety related electronic systems for signalling;
- EN 50159-1: Railway applications - Communication, signalling and processing systems -- Part 1: Safety-related communication in closed transmission systems;
- EN 50159-2: Railway applications - Communication, signalling and processing systems -- Part 2: Safety-related communication in open transmission systems.

CENELEC norms specify the role of the Verification and Validation team, which can be a completely independent company division, of the Assessor, and the techniques and documents needed to guarantee and give evidence of system safety and reliability. In particular, it defines the content of the following documents:

- V&V Plan, containing the planned V&V activities for the system;
- Safety Cases (SC), giving evidence of the performed activities, of the results obtained and of possibly safety application conditions (i.e. procedural prescriptions);
- RAM Analyses, showing models and results of reliability and availability evaluations.

We report a citation from [131]: "A safety case should communicate a clear, comprehensive and defensible argument that a system is acceptably safe to operate in a particular context".

For each critical system, three main types of Safety Cases have to be produced:

- Generic Product
- Generic Application
- Specific Application

The Generic Product Safety Case deals with the validation of the generic HW-SW platform on which the application software runs, comprising system hardware and base software (OS tasks and services, i.e. the lower layers in Figure 5). Such a Safety Case can be maintained unchanged for any new system based on the same platform, or slightly adapted for specific needs.

The Generic Application Safety Case deals with the activities performed to assess application software's safety, comprising integration and functional testing. Generic Application software remains unchanged in any system installation, where only configuration differs. Therefore, it is validated using a specific configuration or a significant set of configurations, until a satisfactory coverage of functionalities is achieved.

The Specific Application Safety Case deals with the safety assurance of a specific configuration, corresponding to a certain system installation. It comprises all the activities needed to verify configuration databases, usually performed using possibly automated static checking approaches.

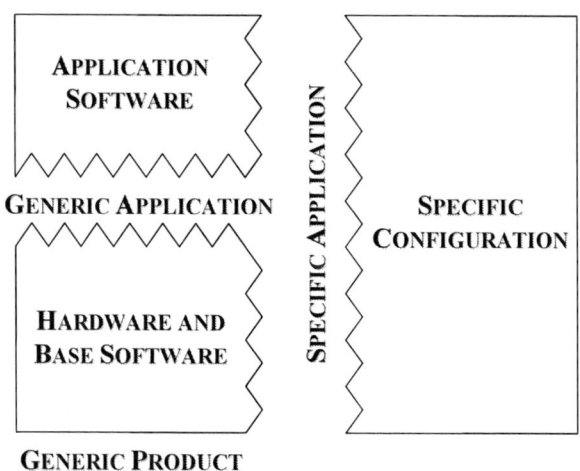

Figure 7. Generic Product, Generic Application and Specific Application.

Figure 7 depicts the relationships between the different part of the system and the related Safety Cases to be produced. As mentioned above, the V&V process is incremental, starting from the base HW-SW platform

(Generic Product SC), passing through application SW (Generic Application SC, which also comprises HW-SW integration tests), and arriving to test the specific installation, constituted by both hardware (sensors and actuators needed by the control system) and configuration database (Specific Application SC, which also comprises Generic Application-Configuration integration testing).

CENELEC norms explicitly advocates and highly recommends (but does not require) the use of formal, semi-formal and structured methods (including CCS, CSP, HOL, LOTOS, OBJ, Temporal Logic, VDM, Z, B and SDL) for developing safety-critical code. In terms of verification procedures, the standard highly recommends, amongst others, formal proofs, finite state machine analysis, and formal methods based inconsistency and correctness analysis. Surprisingly enough, model checking is not one of the recommended verification techniques (see [130]).

2.4. EVALUATION METRICS

SIL-based metrics are both qualitative and quantitative: besides requirements and guidelines described above, in fact, numerical values are given for THR (Total Hazard Rate) and/or MTBHE (Mean Time Between Hazardous Events), usually $< 10^{-9}$ hazardous failures per hour.

Reliability-related indices are instead only quantitative. In the hypothesis of a Homogeneous Poisson Process (HPP) of failure times (which is realistic when the system is in its useful life period, that is no faults due to infant mortality or wear-out are likely to happen), the failure rate λ can be considered as a constant. Therefore, an exponential distribution can be used for failure inter-arrival times, whose mean is defined as Mean Time Between Failures (MTBF). For an exponential distribution, it can be proven that:

$$\lambda = \frac{1}{MTBF} \tag{6}$$

MTBF is by far the most widespread reliability index. When a failure occurs, a time is needed to restore the system to its nominal service. Such a time is known as a Down Time, which is a sum of other elementary times. If we reason in term of means, the Mean Down Time is given by the following expression:

Mean Down Time = Mean Time To Diagnose + Mean Time to Repair + Mean Time to Restart

The time to diagnose represents the time needed to locate the cause of the failure. Restart time also comprises the time for the reintegrated components to perform BIST (Built In Self Tests). Time for the repair team to procure the failed component, move to site and intervene should also be considered, whenever spares are not always availables and system is not supervised on-site 24h/24. However, in practice the MTTR (Mean Time To Repair) index is often used to express the entire down time of the system. The Mean Time To Fail (MTTF), finally, is nothing else than the Mean Time Between Failures minus the Mean Time To Repair, as graphically shown in Figure 8 (where OK stands for system working properly, while KO means system unavailable).

The time system is available (i.e. OK in Figure 8) can be measured in relative terms (%) as follows:

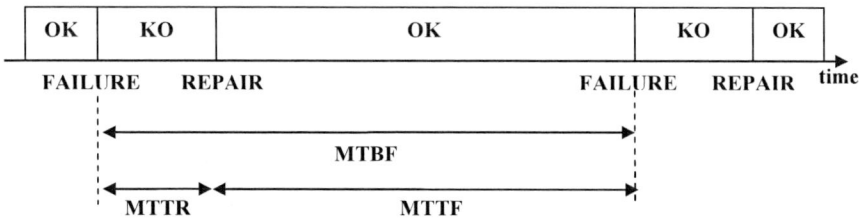

Figure 8. Meaning of reliability related indices.

$$\text{Availability} = \frac{\text{Time system is available}}{\text{Total time}}$$

If system can be modelled by a two states Markov Chain (see Figure 11), then its steady state availability measure can be used, which is often used in system RAMS specification, leading to equation 2..

Therefore, as suggested by intuition, A is a direct consequence of MTBF and MTTR. Unavailability measure U is also used as well, which is simply obtainable as the complement of A to 1:

$$U = 1 - A \tag{7}$$

Typical MTBF for LRU is in the order of 10^5 h in their useful life period, while repair times go from a few minutes to a few hours for highly available systems, according to maintenance strategies and logistics. Simpler components' MTBF can be higher of an order of magnitude, while some typically less reliable components (e.g. power supplies) can feature a lower MBTF.

2.5. METHODOLOGIES AND TOOLS

In order to describe techniques and tools used to assess and ensure system dependability, let us refer to the scheme in Figure 9. The scheme represent what happens (or should happen) in a modern industrial context, where verification and validation involves each phase of system development, since early design (please note that in Figure 9 system specification and design have been collapsed in the same phase, as the output of both steps are of the same type, namely documents of requirements). The phases are variously interconnected, as at each step of system development it is necessary to predict the effect of the produced output on system dependability, actuate possible corrective actions whenever problems are detected, and then verify that the modifications did not have a negative impact on other parts of the system. Furthermore, each step of the dependability assurance cycle can exploit data coming from following development phases, whenever they have been already performed for the same or similar systems.

Starting from the scheme of Figure 9, we will shortly present in the following the verification and validation techniques which most widespread in an industrial context. A brief survey of safety management tools and techniques can be found in [117].

In the design phase, dependability prediction is performed by means of two main classes of techniques:

- Manual techniques, e.g. specification revision, requirements classification, traceability analysis, etc.
- Automatic techniques, e.g. formal methods, which can be used after a model has been built, starting from the requirements and using a suitable language.

Such techniques are meant to:

- Find specification incompleteness, incorrectness, ambiguities and incoherencies in terms of:
  - Qualitative or functional aspects;

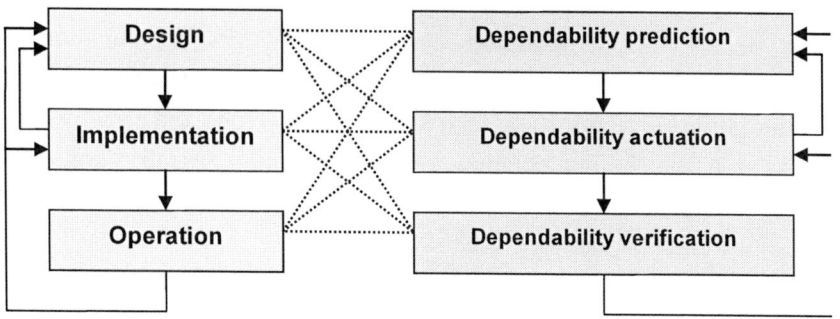

Figure 9. Scheme of a possible phase-independent dependability assurance cycle.

  - Quantitative or non functional aspects (e.g. unmatched availability requirements between system and constituents).

- Size system architecture (e.g. redundancy of fault-tolerance mechanisms) and fine tune reliability parameters in order to meet system requirements since early design stages.

Widespread natural language requirements management tools are Rational Requisite Pro [14] and Telelogic Doors [15].

Most widespread languages for real-time systems' specification are B and Z. First Order and Temporal Logic, Timed Automata, Stochastic Petri Nets, Binary Decision Diagrams are examples of formal languages used to model and check natural language specification [12]. Most widespread techniques for specification and design validation are model-checking and theorem proving, usually applied to software, e.g. using languages like Promela together with the SPIN tool [139]. Quantitative reliability and maintainability analyses are usually applied to hardware and performed by means of Fault Trees, Markov Chains and, more recently, Bayesian Networks (data from the field are used whenever available in order to size model parameters). The Unified Modeling Languages has also been employed in order to verify real-time systems (see [13] for an application to a critical communication protocol). A promising approach for the early stage dependability prediction based on a UML specification was introduced in the HIDE project [23]. Each technique features advantages and limitations in terms of expressive power, efficiency and easy of use.

As for the implementation phase, prediction usually consists in checking software coding or hardware construction. For hardware, stress tests and fault injection are used. For software, we can list two types of analysis:

- Static analysis, aimed at checking code quality, software architecture and test coverage;
- Dynamic analysis, best known as software testing.

The first technique is usually supported by static analysis tools who automatically:

- Check for predefined conditions (e.g. function length in Lines Of Code metric);
- Show software architecture in term of modules and function call graphs;
- Measure code coverage, e.g. using the Decision to Decision Path metric (a kind of decision coverage).

An industrially widespread tool supporting many of such analysis is Telelogic Logicscope [148].

Dynamic analyses are module, integration and system testing. Module and integration testing can be supported by commercial tools (e.g. Cantata [93]). System testing is instead supported by application specific simulation environments by means of usually proprietary approaches, as it is described in more details in [68].

After an implementation flaw is detected, a notification (known as System Problem Report, SPR) is usually formalized by the V&V division and sent to the Development Division, which will take in charge the modification. The last verification step for software consists in Regression Testing, aimed at checking that modifications have been correctly implemented with no negative impact on other functionalities.

In the operational phase, dependability prediction consists in diagnosing in anticipation potential cause of problems, by means of specific hardware probes and software diagnostic tasks. Quantitative problems are instead statistically detected by means of FRACAS results' analysis. Whenever a "qualitative" problem is detected, dependability actuation consists in preventive maintenance or system shut-down (the latter to be performed in most critical cases). A design review is instead performed when on-the-field data reveal not coherent with statistical prediction on system reliability parameters. Finally, operational dependability verification requires performing a set of built-in self tests (in case of hot reintegration of a single component) and/or start-up tests (in case of system shutdown); in the "quantitative case", verification just means that new design choices have to be checked for correctness by the usual means but using the new available on-the-field component reliability data (which lead to more realistic results).

The main dependability prediction techniques are described in detail in the following sections.

Chapter 3

DEPENDABILITY PREDICTION TECHNIQUES

As aforementioned, dependability prediction techniques can be classified into two main classes: simulative and formal approaches. They both can applied to hardware and software layers, both for availability and safety evaluation, but while formal methods can be applied since early design stages, simulation based techniques need at least a prototype of the system in order to be employed. In industrial practice, basic formal methods (e.g. Fault Trees [128]) and are widespread for structural reliability studies, while simulative approaches are largely used for functional safety assurance. When an abstraction is produced of the system whose dependability need to be predicted, then we can speak about model-based approaches. Models are mandatory when using formal approaches and can be highly advantageous also with simulation.

3.1. SIMULATIVE TECHNIQUES FOR FUNCTIONAL SAFETY EVALUATION

Simulation of the system or of a part of it is possible when the following requirements are met:

- A (early) prototype or model of the hardware/software system is available;
- A simulation environment interfacing with the real system has been developed.

The first requirement is obvious: we need something to simulate, whether a hardware component, an executable piece of code or the entire system. The second requirement is needed in order to artificially recreate the stimuli coming from the external world. This is often achieved by designing proper "stubs" (i.e. simplified software models) representing the interfaces of the interacting entities.

The most important distinction is whether the real hardware is used or a simulated version, e.g. using a commercial PC instead of the target embedded hardware. According to the aim of the simulation, different solutions are possible. For instance, when evaluating the correctness of high level system logic, just the real code is necessary, and it is not needed to be executed on the real hardware; when performing hardware-software integration testing or physical fault-injection, a system-in-the-loop scheme is needed, in which simulators interact with the real system under test (i.e. the "target" system).

At the hardware verification level, simulation is often intended as a synonym of fault-injection[1], which can be:

- Simulated, that is performed on a HDL model of the real system;
- Physical, that is performed on the target system.

The first one features several advantages in terms of flexibility, as input domains and fault models can be freely selectable. The second one need a real hardware prototype (which can be available only at later development stages), but of course allows for more realistic results. In both cases, statistical analyses can also be performed by means of accelerated simulations. Sometimes also the so called Measurement Based Analysis is

---

[1] Refer to [115] for a survey of fault injection techniques and available tools.

listed among the hardware simulation techniques, in which real on-the-field data is registered, filtered and analyzed; however, such a technique, like the mentioned FRACAS, is only employable during the operational life of the system. Simulation environment for physical fault-injection are not in the scope of this work; for a related description the reader can refer to [1], describing the LIVE approach and toolset.

At the software verification level, simulation is a synonym of testing. Testing is the most complex and time consuming activities within the development of dependable real-time systems [156]. A test-case is simply an (INPUT SEQUENCE - EXPECTED OUTPUT) combination at system's interface[2]. A test-suite is a set of test-cases grouped by some property. In particular, two main testing classes can be defined:

- White-box (i.e. glass-box) or structural testing;
- Black-box or functional testing.

The first one is performed by accessing the internal part of the software, knowing its structure and low-level behavior. Module and module integration testing are examples of white box testing, in which one or more software modules (e.g functions) are tested in order to evaluate that they give correct results for any possible class of inputs and they interact properly. Inputs can be injected at the function level or at system level. Flow-graph based techniques, representing the order of execution of statements or the evolution of values of variables, are widespread for structural testing [82]. Several tools have been developed to support structural testing (see e.g. [24]). The code coverage measure gives to test engineers a feedback of how much code has been exercised by the executed tests; whenever a significant part of the code is not covered, ad-hoc integrative tests have to be designed and executed, after having analyzed the uncovered piece of software. To perform module testing, a verifier must know low level software specification, which is often provided in code documentation (e.g. SDD, function headers, in code comments), and it is very useful to read and understand code structure and behavior (by means of code inspection), while he/she has not to know high level system functional requirements.

The second kind of testing is instead focused on system functional requirements. Testing is meant to cover requirements by executing proper input sequences and reading corresponding outputs at system's interface. The definition of expected output is performed by a so called "oracle", which can be human or given by a parallel independent model[3]. The definition of test-cases is not straightforward, as the coverage of a functional requirement can require the execution of a high number of tests, according to boundary analysis, worst case testing and robustness checking approaches performed on input classes [82]. Exhaustive black-box testing can be proven to be impossible to achieve, but combining white and black box testing can improve both test effectiveness and efficiency. This is called grey-box testing and it is particularly important for complex critical systems, for which it is essential:

- Not to miss any significant input sequence in the functional testing phase, for obvious safety reasons;
- To accurately select the minimum set of input sequences required to stimulate all system functionalities, in order to contain the number of required test-cases, which could easily explode when combining inputs in any possible way.

Grey-box testing allows to access system internal structure in order to e.g. accurately select the test-cases to be executed or to measure test effectiveness by means of the code coverage measure; it is particularly suited when performing HW-SW integration or system tests, as such tests are the most critical in terms of budget and results.

---

[2] In practice, the definition of a test-case also includes an identifier and possibly environmental and procedural requirements (sometimes known as "preconditions").
[3] Despite of its simple definition, developing an adequate automatic oracle is an error prone and costly activity.

3.2. FORMAL TECHNIQUES FOR STRUCTURAL AVAILABILITY EVALUATION

"Applied to computer systems development, formal methods provide mathematically based techniques that describe system properties. As such, they present a framework for systematically specifying, developing, and verifying systems." (citation from [97]). Formal methods are employed in a variety of industrial applications, from microprocessor design to software engineering and verification (see [12] for a survey of most widespread methods and their successful applications).

Despite of such variety of methods and applications, we will concentrate on the application of formal methods in structural availability evaluation, and thus we will restrict our attention to the related suitable formal languages. The motivation is that at the state of the art, no formal method seems suitable to represent an entire system from both a structural and functional point of view. Limitations regard expressive power or complexity of the solving algorithms. While it is possible to find methods with which hardware complexity can be managed, at least when modelling dependability related aspects, extensive formal modelling of system level functional aspects seems still unfeasible using nowadays available tools. This is the reason why well established formal methods are largely employed in industrial context only for reliability analyses. We will shortly present advantages and limitations of such methods in the following of this section. Multiformalism techniques, combining more than one formalism in a single model, have still a very limited diffusion in industrial practice.

3.2.1. Fault Trees

The Fault Tree (FT) formalism is widespread for structural reliability, availability and safety modeling. It links failure causes (Basic Events, BE) to consequence (Top Event, TE) using a tree structure and Boolean connectors, following a deductive approach (see Figure 10).

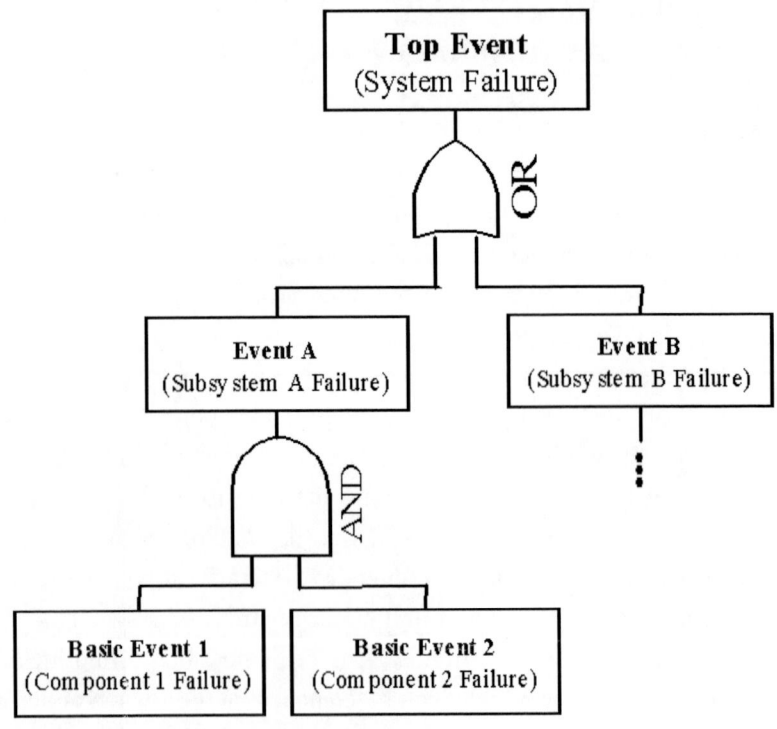

Figure 10. An example Fault Tree.

It has been developed and used for system reliability evaluation of chemical and nuclear plants [150], and it is nowadays applied to a great variety of computer based control systems.

Its main advantages are the ease of use, as it is a graphical and intuitive formalism, and efficiency, as solving algorithms are based on combinatorial techniques (i.e. minimal cut sets).

The limitation consists in expressive power: for its simplicity, it does not allow to evaluate articulate repair strategies and other complex behaviors. Furthermore, events must be statistically independent, as it does not allow to model dependencies. When used for availability evaluation, the only maintenance assumption is perfect repair.

To cope with such limitations, several extensions have been proposed, comprising Parametric Fault Trees (PFT) [18], Dynamic Fault Trees (DFT)

[96] and Repairable Fault Trees [51], which are solved by prior translation into Stochastic Petri Nets).

3.2.2. Reliability Block Diagrams

The Reliability Block Diagram (RBD) formalism is one of the most widespread technique for system reliability modeling. Subsystems of components are graphically represented by blocks, connected in series and parallel structures, following an inductive approach (from causes to consequences) [150].

Its applications, advantages and limitations are the same of FT, as it features the same expressive power and solving efficiency.

3.2.3. Continuous Time Markov Chains

The Continuous Time Markov Chain (CTMC) formalism is quite widespread for availability modeling of repairable systems [5]. It features better expressive power but worse efficiency with respect to FT and RBD, being state based.

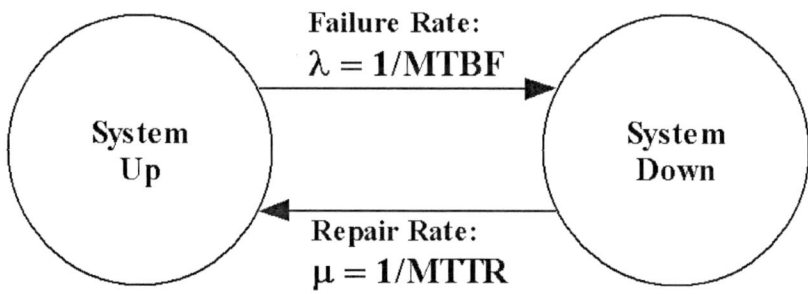

Figure 11. The simplest CTMC model for reliability modelling.

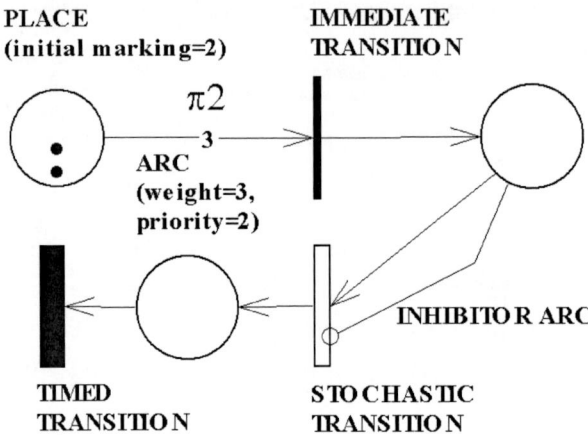

Figure 12. GSPN semiotic.

It is quite ease to use and allows to model non banal repair strategies (but not any); however, suffering from the state space explosion problem, it is not suited to model complex systems.

3.2.4. Stochastic Petri Nets

The Petri Net (PN) formalism is a general purpose graphical specification language, widespread for modeling concurrent software systems. Its generalized stochastic extension (Generalized Stochastic Petri Nets, GSPN[4]) can be used for dependability analyses of both structural and behavioral aspects (including performability), and has been employed in a variety of applications (see Figure 12 for formalism semiotic; an in depth description of its syntax and semantic is provided in [112]). The main advantage of GSPN is in its expressive power, while the main limitation consists in the low efficiency of the solving algorithms, which are state-based; furthermore, for an advanced use they require a skilled modeler. Stochastic Well-formed Nets (SWN) has been introduced to reduce model

---

[4] The GSPN formalism merges Timed Petri Nets (introducing timed transitions, which fire after a specified time) and Stochastic Petri Nets (introducing stochastic transitions, whose firing time is distributed as an exponential random variable).

complexity by folding that is by exploiting model symmetries [73]. Colored Petri Nets (CPN) is another PN extension providing enhanced features which make them more similar to general purpose programming languages [108]. Both in SWN and CPN tokens are no more indistinguishable and they are assigned a semantic, represented by a color class.

3.2.5 Bayesian Networks

The Bayesian Network (BN) formalism is one of the languages suited to model uncertainty, traditionally used in artificial intelligence applications [64]. A BN is DAG (Direct Acyclic Graph), constituted by places, representing stochastic variables, and arcs, representing statistical dependencies (by means of Conditional Probabilities Tables, CPT) between such variables (see Figure 13). Solving a BN means detecting a posteriori probability from a priori ones (the "evidence"), or viceversa, exploiting the Bayes theorem result. Recently, BN have been successfully applied to dependability modeling, both of software quality (see [63] and [106]) and system reliability (see [19]). Their solving algorithms can be proven to be NP-hard; however, being non-state based, they feature better efficiency with respect to GSPN, at least in structural dependability evaluation (see [21]). BN models are quite easy to build, but in general modelers must be very careful to correctly translate reality into model features; such translation is less error prone for structural models.

The Dynamic Bayesian Network (DBN) formalism is a BN extension that allows modelers to specify time-varying versions of the network, thus introducing the concepts of time dependency and state of the network.

BN can be extended to decision networks by adding decision and utility nodes, which can be used for the optimization of design choices (which are assigned a cost).

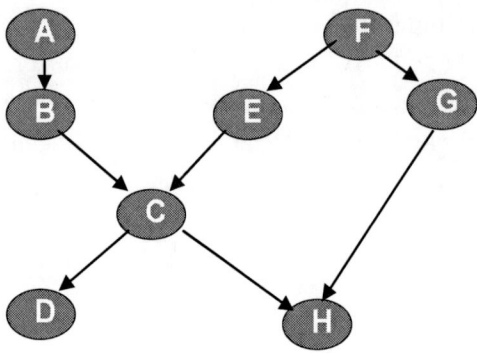

Figure 13. The graphical representation of a Bayesian Network.

3.2.6 Solvers

A number of tools, GUIs and solvers are available to manage the building and solution of models specified using the aforementioned graphical formalisms. We will concentrate, whenever possible, on freeware and open source software applications provided by academic research centers; while their advantages are obvious, possible disadvantages consist in limited stability and maintainability. SHARPE [134] is one of these tools, and it is able to solve FT, RBD, CTMC and GSPN models; however, it does not feature a GUI, therefore models must be expressed in a textual form. GreatSPN [74] is among the most widespread graphical tools for GSPN analysis, featuring both analytical and simulation based solving algorithms, providing transient and stead-state solutions, and model checking capabilities, which are very useful to verify important net properties like ergodicity and liveliness. Even though freeware and open source GUI based modeling environments for BN exist (e.g. Bayesian Networks in Java, BNJ [40]; see [109] for a quite complete survey), featuring numerous types of solving algorithms, commercial tools like Netica by Norsys [125] and Hugin Expert [89] seem to be much more stable and complete.

3.2.7. Comparison of Structural Reliability Modeling Techniques

The choice of the most suited formalism for structural dependability evaluation depends on several factors, among which:

- The familiarity of modeler with a given formalism (related to knowledge, skill, experience);
- The modeling power of the formalism: is the formalism able to model (with simplicity) every aspect of interest?
- Efficiency and scalability: is the formalism able to deal with large systems?
- Reusability and composability: how much flexibility the formalism and related tools offer to the modeler?

As a practical rule, the simplest formalism able to satisfy modeler's needs should be employed. A comparison between formalisms (partly taken from [21]), with respect to modeling power and system dimensions, has been reported in Figure 14: it can be noted that BN can be a good compromise between FT and GSPN when just a limited extension to expressive power is needed, e.g. to model statistical dependencies between components.

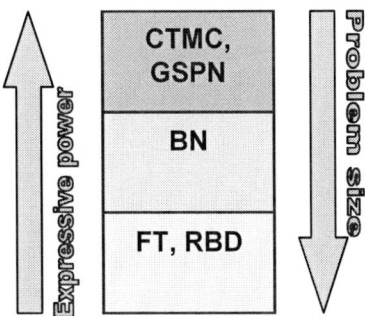

Figure 14. Comparison of different reliability modelling formalisms.

GLOSSARY OF ACRONYMS

A list of the acronyms used more often in the text follows.

Abbreviation	Meaning
ABS	Automatic Braking System
ALARP	As Low As Reasonably Practicable
ATC	Automatic Train Control or Air Traffic Control
ATM	Automatic Transaction Module
ATP	Automatic Train Protection
BA	Boundary Analysis
BDD	Binary Decision Diagram
BE	Basic Event
BIST	Built In Self Test
BN	Bayesian Network
BTM	Balise Transmission Module
COTS	Commercial Off The Shelf
CPN	Coloured Petri Nets
CPT	Conditional Probability Table
CPU	Central Processing Unit
CTMC	Continuous Time Markov Chain
CTP	Change of Traction Power
DAG	Direct Acyclic Graph
DBN	Dynamic Bayesian Network

DDP	Decision to Decision Path
DFT	Dynamic Fault Trees
DMI	Driver Machine Interface
EDRM	Error Detection and Recovery Mechanism
ERTMS/ETCS	European Railway Traffic Management System / European Train Control System
EVC	European Vital Computer
FIS	Functional Interface Specification
FME(C)A	Failure Mode Effects (and Criticality) Analysis
FPGA	Field Programmable Gate Array
FRACAS	Failure Reporting Analysis and Corrective Action System
FRS	Functional Requirements Specification
FSM	Finite State Machine
FT	Fault Tree
FTA	Fault Tree Analysis
GSPN	Generalized Stochastic Petri Nets
GUARDS	Generic Upgradeable Architecture for Real-Time Dependable Systems
GUI	Graphical User Interface
HA	Hazard Analysis
HDL	Hardware Description Language
HDS	Hardware Design Specification
HIDE	High-level Integrated Design Environment for Dependability
HMI	Human Machine Interface
HPP	Homogeneous Poisson Process
HW	Hardware
ISO/OSI	International Standards Organization / Open Systems Interconnection
JRU	Juridical Recording Unit
LCSAJ	Linear Code Sequence and Jump
LIVE	Low-Intrusion Validation Environment
LOC	Lines of Code

Glossary of Acronyms

LRU	Legal Recording Unit
LTM	Loop Transmission Module
MA	Movement Authority
MC	Markov Chain
MCS	Monte Carlo Simulation
MDT	Mean Down Time
MEM	Minimal Endogen Mortality
MMI	Man Machine Interface
MSC	Message Sequence Chart
MTBF	Mean Time Between Failures
MTBHE	Mean Time Between Hazardous Events
MTTD	Mean Time To Diagnose
MTTF	Mean Time To Fail
MTTR	Mean Time To Repair
OMG	Object Management Group
OS	Operating System
PC	Personal Computer
PFT	Parametric Fault Trees
PLC	Programmable Logic Controller
QN	Queuing Network
RAID	Redundant Array of Independent Disks
RAMS	Reliability Availability Maintainability Safety
RB	Repair Box
RBC	Radio Block Center
RBD	Reliability Block Diagrams
RFT	Repairable Fault Trees
RT	Real Time
RTM	Radio Transmission Module
SC	Safety Case
SDD	Software Design Description
SECT	Strong Equivalence Class Testing
SIL	Safety Integrity Level
SPN	Stochastic Petri Net
SPR	System Problem Report

SRS	System Requirements Specification
SSRS	Sub-System Requirements Specification
STM	Specific Transmission Module
SW	Software
SWN	Stochastic Well-formed Nets
SWRS	Software Requirements Specification
TA	Timed Automata
TC	Track Circuit
TE	Top Event
THR	Total Hazard Rate
TIU	Train Interface Unit
TMR	Triple Modular Redundancy
UML	Unified Modeling Language
UPS	Uninterruptible Power Supply
V&V	Verification and Validation
WCT	Worst Case Testing
WFE	Workflow Engine
XML	Extended Markup Language

REFERENCES

[1] Martin L. Shooman: Reliability of Computer Systems and Networks: Fault Tolerance, Analysis, and Design. John Wiley & Sons Inc., 2002.

[2] D. Harel, A. Pnueli: *On the development of reactive systems*. In Logics and Models of Concurrent Systems, volume F-13 of NATO Advanced Summer Institutes, pages 477--498. Springer-Verlag, 1985.

[3] Robin E. McDermott et al., *The Basics of FMEA*, Resource Engineering Inc., 1996.

[4] A. Avizienis, J. C. Laprie, B. Randel, Fundamental Concepts of Dependability, LAAS Report n. 01-145, 2001.

[5] H. Ascher and H. Feingold, Repairable Systems Reliability, Marcel Dekker, Inc., New York, 1984.

[6] M. Lyu, *Software Fault Tolerance,* John Wiley & Sons, 1995.

[7] VITA Web Site, VME Page: http://www.vita.com/learn.html

[8] Wind River Web Site, VxWorks Page: http://www.windriver.com/products/run-time_technologies/Real-Time_Operating_Systems/VxWorks_6x/

[9] OMG Model Driven Architecture Home Page: http://www.omg.org/mda/

[10] Kececi Nihal, Mohammad Modarres: *Software Development Life Cycle Model to Ensure Software Quality*. International Conference on Probabilistic Safety Assessment and Management, New York City, New York, September 13-18, 1998.

[11] P. di Tommaso, R. Esposito, P. Marmo, A. Orazzo: Hazard Analysis of Complex Distributed Railway Systems. In *Proceedings of 22nd International Symposium on Reliable Distributed Systems,* Florence (2003) 283-292.

[12] Edmund M. Clarke, Jeannette M. Wing: Formal Methods: State of the Art and Future Directions. *ACM Comput. Surv. 28*(4): 626-643 (1996).

[13] R. Esposito, A. Sanseviero, A. Lazzaro, P. Marmo: Formal Verification of ERTMS Euroradio Safety Critical Protocol. In *Proceedings of FORMS 2003*, May 15-16, 2003, Budapest, Hungary.

[14] IBM Rational Requisite Pro Home Page: http://www-306.ibm.com/software/awdtools/reqpro/

[15] Telelogic Doors Home Page: http://www.telelogic.com/corp/products/doors/index.cfm

[16] A. Amendola, P. Marmo, and F. Poli. Experimental Evaluation of Computer-Based Railway Control Systems. In *Symp. on Fault-Tolerant Computing* (FTCS-97), pages 380--384, June 1997.

[17] A. Bobbio, D. Codetta Reiteri: Parametric Fault-trees with dynamic gates and repair boxes. In *Proc. Reliability and Maintainability Symposium,* Los Angeles, CA USA, Jan. 2004, pp. 459-465.

[18] A. Bobbio, G. Franceschinis, R. Gaeta, G. Portinaie: Parametric fault tree for the dependability analysis of redundant systems and its high-level Petri net semantics. In *IEEE Transactions on Software Engineering,* vol. 29 issue 3, March 3rd 2003, pp. 270-287.

[19] A. Bobbio, L. Portinale, M. Minichino, E. Ciancamerla, Improving the Analysis of Dependable Systems by Mapping Fault Trees into Bayesian Networks, *Reliability Engineering and System Safety Journal* – 71/3 – pp 249-260, 2001.

[20] A. Bobbio, L. Portinale, M. Minichino, E. Ciancamerla: "Improving the Analysis of Dependable Systems by Mapping Fault Trees into Bayesian Networks". In *Reliability Engineering and System Safety Journal* – 71/3 – pp 249-260, 2001.

[21] A. Bobbio, S. Bologna, E. Ciancamerla, G. Franceschinis, R. Gaeta, M. Minichino, L. Portinale, Comparison of Methodologies for the Safety and Dependability Assessment of an Industrial Programmable Logic Controller. In *Proc. 12th European Safety and Reliability Conference* (ESREL 2001), 16-20 Settembre 2001, Torino, Italy.

[22] A. Bondavalli et al. Design Validation of Embedded Dependable Systems. In *IEEE MICRO,* September-October 2001, 52-62.

[23] A. Bondavalli, M. Cin, D. Latella and A. Pataricza: *High-level Integrated Design Environment for Dependability*. In *Proc. Fifth Int. Workshop on Object-Oriented Real-Time Dependable Systems (WORDS-99),* Monterey, California, USA, November 18-20, 1999.

[24] A. Caiazza, R. Di Maio, F. Scalabrini, F. Poli, L. Impagliazzo, A. M. Amendola: A New Methodology and Tool Set to Execute Software Test on Real-Time Safety-Critical Systems. In: Lecture Notes in Computer Science (LNCS) Vol. 3463 (ed. Springer-Verlag Heidelberg): *The Fifth European Dependable Computing Conference, EDCC-5, Budapest, Hungary, April* 20-22, 2005: pp. 293-304.

[25] A. Chiappini, A. Cimatti, C. Porzia, G.Rotondo, R. Sebastiani, P.Traverso, A.Villafiorita. Formal Specification and Development of a Safety-Critical Train Management System. In *Proceedings of the 18th International Conference on Computer Safety,* Reliability and Security (SAFECOMP'99). Toulouse, FRANCE. September 1999. Lecture Notes in Computer Science series (LNCS) 1698, pages 410-419.

[26] A. Cimatti, F. Giunchiglia, G. Mongardi, D. Romano, F. Torielli, P. Traverso: Formal Verification of a Railway Interlocking System using Model Checking. *Formal Aspects of Computing* 10(4): 361-380 (1998).

[27] A. Egyed, N. Medvidovic: Consistent Architectural Refinement and Evolution using the Unified Modeling Language. In *Proceedings of 1st Workshop on Describing Software Architecture with UML*, co-located with ICSE 2001.

[28] A. M. Amendola, L. Impagliazzo, P. Marmo, G. Mongardi and G. Sartore, "Architecture and Safety Requirements of the ACC Railway Interlocking System", in *Proc. 2nd Annual Int. Computer Performance & Dependability Symposium* (IPDS'96), (UrbanaChampaign, IL, USA), pp.21-29, IEEE Computer Society Press, September 1996.

[29] A. Zimmermann, G. Hommel. Toward Modeling and Evaluation of ETCS Real-Time Communication and Operation. *Journal of Systems and Software archive* Vol 77(1) Special issue: Parallel and distributed real-time systems. Pages: 47 – 54. ISSN:0164-1212. 2005. Elsevier Science Inc. New York, NY, USA.

[30] Almeida Jr., J. R., Camargo Jr., J. B., Basseto, B. A., and Paz, S. M. 2003. Best Practices in Code Inspection for Safety-Critical Software. In IEEE Software 20, 3 (May. 2003), 56-63.

[31] Anand and A.K. Somani: Hierarchical analysis of fault trees with dependencies, using decomposition. In Proc. Annual Reliability and Maintainability Symposium, 1998, pp. 69-75.

[32] Andrea Bondavalli, Mario Dal Cin, Diego Latella, István Majzik, András Pataricza, Giancarlo Savoia: *Dependability analysis in the early phases of UML-based system design*. In Computer Systems Science and Engineering, 16(5), 2001, pp. 265-275.

[33] ANSI/IEEE Std 1012-1986, *"IEEE Standard for Software Verification and Validation Plans,"* The Institute of Electrical and Electronics Engineers, Inc., February 10, 1987.

[34] ANSI/IEEE Std 730-1984, *"Standard for Software Quality Assurance Plans,"* The Institute of Electrical and Electronics Engineers, Inc., 1984.

[35] B. Dugan, K. J. Sullivan, D. Coppit: Developing a Low-Cost High-Quality Software Tool for Dynamic Fault-Tree Analysis. In *IEEE Transactions on Reliability,* vol. 29, 2000, pp. 49-59.

References

[36] B. Jeng, E.J. Weyuker: Some Observations on Partition Testing. In Proceedings of the ACM SIGSOFT '89 *Third Symposium on Software Testing,* Analysis and Verification, Key West (1989).

[37] Beizer, Boris, *"Software Testing Techniques",* 2nd edition, New York: Van Nostrand Reinhold, 1990.

[38] Bernardi, S., Donatelli, S., and Merseguer, J.: From UML Sequence Diagrams and Statecharts to Analysable Petri Net models. In *Proceedings of the 3rd international Workshop on Software and Performance* (WOSP'02), Rome, Italy, July 24 - 26, 2002: pp. 35-45.

[39] Bernardi, S., Donatelli, S., Building Petri net scenarios for dependable automation systems, In Proc. of the 10th Int. Workshop on Petri nets and performance models (PNPM03), Urbana-Champaign, IL, USA, September 2--5 2003. *IEEE Comp. Soc. Press.* Pages: 72 – 81.

[40] BNJ Home Page: W. H. Hsu, R. Joehanes, Bayesian Network Tools in Java (BNJ) v2.0, http://bndev.sourceforge.net

[41] Bred Pettichord: Success with Test Automation. In *Quality Week,* San Francisco (2001).

[42] C. Abbaneo, F. Flammini, A. Lazzaro, P. Marmo, A. Sanseviero: UML Based Reverse Engineering for the Verification of Railway Control Logics. In: *Proceedings of Dependability of Computer Systems* (DepCoS'96), Szklarska Poręba, Poland, May 25-27, 2006: pp. 3-10.

[43] C. Bernardeschi, A. Fantechi, S. Gnesi, S. La Rosa, G. Mongardi, D. Romano. A Formal Verification Environment for Railway Signalling System Design. Formal Methods In System Design. *ISSN* 0925-9856. Vol 12:139-162. 1998.

[44] C.R. Cassady, E.A. Pohl, W.P. Murdock. Selective Maintenance Modeling for Industrial Systems. In *Journal of Quality in Maintenance Engineering,* Vol. 7, No. 2, 2001: pp. 104-117.

[45] CENELEC EN 50126: *Railway applications - The specification and demonstration of Reliability, Availability, Maintainability and Safety* (RAMS), 2001.

[46] CENELEC EN 50128: *Railway Applications - Communication, signalling and processing systems - Software for railway control and protection systems*, 2001.
[47] CENELEC EN 50129: *Railway applications - Communication, signalling and processing systems - Safety related electronic systems for signalling*, 2003.
[48] CENELEC EN 50159-1: Railway applications - Communication, signalling and processing systems -- Part 1: *Safety-related communication in closed transmission systems*, 2001.
[49] CENELEC EN 50159-2: Railway applications - Communication, signalling and processing systems -- Part 2: *Safety-related communication in open transmission systems*, 2001.
[50] CENELEC Home Page: https://www.cenelec.org
[51] Codetta Raiteri, D., Franceschinis, G., Iacono, M., Vittorini, V., Repairable Fault Tree for the Automatic Evaluation of Repair Policies, In *IEEE Proceedings of the International Conference on Dependable Systems and Networks* (DSN'04), Florence, Italy, June 28-July 1, 2004: p. 659.
[52] Cung A., Lee Y.S.: Reverse Software Engineering with UML for website maintenance. In *IEEE Proceedings of Working Conference in Reverse Engineering '00*, pp. 100-111 (2000).
[53] D. Bjørne. The FME Rail Annotated Rail Bibliography. http://citeseer.ist.psu.edu/279277.html
[54] D. P. Siewiorek, R. Swarz, R*eliable Computer Systems: Design and Evaluation,* 3rd ed., A.K. Petres, Ltd., 1998.
[55] D. Powell et al., *A Generic Fault-Tolerant Architecture for Real-Time Dependable Systems,* Kluwer Academic Publishers, 2001.
[56] D.R. Wallace and R.U. Fujii, Eds., *IEEE Software: Special Issue on Software Verification and Validation,* IEEE Computer Society Press, May, 1989.
[57] Daniel Duane Deavours, *Formal Specification of the Möbius Modeling Framework.* Ph.D. Thesis, University of Illinois at Urbana-Champaign, 2001.

[58] Dave Astels: Refactoring with UML. In *Proc. of the 3rd International Conference on eXtreme Programming and Flexible Processes in Software Engineering*, 2002: pp. 67-70.

[59] E. Dustin, J. Rashka, J. Paul: *Automated Software Testing*, Addison Wesley (1999).

[60] E. J. Chikofsky, J. H. Cross: *Reverse Engineering and Design Recovery:* A Taxonomy. In IEEE Software, vol. 7, no. 1, January 1990.

[61] E.A. Pohl, E.F. Mykytka. Simulation Modeling for Reliability Analysis. In *Tutorial Notes of the Annual Reliability & Maintainability Symposium*, 2000.

[62] Egyed, N. Medvidovic: Consistent Architectural Refinement and Evolution using the Unified Modeling Language. In *Proceedings of ICSE 2001*, Toronto, Canada, May 2001, pp. 83-87.

[63] ENEA ISA-EUNET Presentation: http://tisgi.casaccia.enea.it/ projects/isaeunet/SafetyCase/ppframe.htm

[64] Eugene Charniak, *Bayesian Networks without Tears*, AI Magazine, 1991.

[65] F. Bause, P. Buchholz, and P. Kemper. *A toolbox for functional and quantitative analysis of DEDS*. In *Proc. 10th Int. Conf. on Modelling Techniques and Tools for Computer Performance Evaluation*, Lecture Notes in Computer Science 1469, Springer-Verlag, 1998: pp. 356-359.

[66] F. Flammini, M. Iacono, S. Marrone, N. Mazzocca: "Using Repairable Fault Trees for the evaluation of design choices for critical repairable systems". In: *Proceedings of the 9th IEEE International Symposium on High Assurance Systems Engineering* (HASE2005), Heidelberg, Germany, October 12-14, 2005: pp. 163-172.

[67] F. Flammini, M. Iacono, S. Marrone, N. Mazzocca: Using Repairable Fault Trees for the evaluation of design choices for critical repairable systems. In: *Proceedings of the 9th IEEE International Symposium on High Assurance Systems Engineering*

(HASE2005), Heidelberg, Germany, October 12-14, 2005: pp.163-172.

[68] F. Flammini, P. di Tommaso, A. Lazzaro, R. Pellecchia, A. Sanseviero: The Simulation of Anomalies in the Functional Testing of the ERTMS/ETCS Trackside System. In: *Proceedings of the 9th IEEE International Symposium on High Assurance Systems Engineering* (HASE2005), Heidelberg, Germany, October 12-14, 2005: pp.131-139.

[69] F. Flammini, S. Marrone, N. Mazzocca, V. Vittorini: Modelling Structural Reliability Aspects of ERTMS/ETCS by Fault Trees and Bayesian Networks. In *Proceedings of the European Safety and Reliability Conference,* ESREL 2006, Estoril, Portugal, September 18-22, 2006.

[70] F. Moscato, N. Mazzocca, V. Vittorini: "Workflow Principles Applied to Multi-Solution Analysis of Dependable Distributed Systems". In *Proceedings of 12th Euromico Conference on Parallel, Distributed and Network-Based Processing* (PDP'04), 2004, p.134

[71] Francesco Moscato, Marco Gribaudo, Nicola Mazzocca and Valeria Vittorini, Multisolution of complex performability models in the OsMoSys/DrawNET framework. In *IEEE Proceedings of the 2nd International Conference on Quantitative Evaluation of Systems* (QEST'05), 2005.

[72] Francesco Moscato: *"Multisolution of Multiformalism Models: Formal Specification of the OsMoSys Framework".* PhD thesis, Second University of Naples, October 2005.

[73] G. Chiola, C. Dutheillet, G. Franceschinis, S. Haddad: *Stochastic Well-Formed Colored Nets and Symmetric Modeling Applications.* In *IEEE Transactions on Computers,* vol. 42, 1993, pp. 1343-1360.

[74] G. Chiola, G. Franceschinis, R. Gaeta, and M. Gribaudo, GreatSPN 1.7: Graphical Editor and Analyzer for Timed and Stochastic Petri Nets. *Performance Evaluation, special issue on Performance Modeling Tools,* 24(1&2): 47-68, November 1995.

[75] G. Ciardo, A. Miner: *SMART: Simulation and Markovian Analyser for Reliability and Timing.* In *Proc. 2nd International Computer*

Performance and Dependability Symposium (IPDS'96), IEEE Computer Society Press, 1996, page 60.

[76] G. De Nicola, P. di Tommaso, R. Esposito, F. Flammini, A. Orazzo: A Hybrid Testing Methodology for Railway Control Systems. In: *Lecture Notes in Computer Science (LNCS)* Vol. 3219 (ed. Springer-Verlag Heidelberg): Computer Safety, Reliability, and Security: 23rd International Conference, SAFECOMP 2004, Potsdam, Germany, September 21-24, 2004: pp.116-135.

[77] G. De Nicola, P. di Tommaso, R. Esposito, F. Flammini, P. Marmo, A. Orazzo: ERTMS/ETCS: Working Principles and Validation. In: *Proceedings of the International Conference on Ship Propulsion and Railway Traction Systems*, SPRTS 2005, Bologna, Italy, October 4-6, 2005: pp.59-68.

[78] G. De Nicola, P. di Tommaso, R. Esposito, F. Flammini, P. Marmo, A. Orazzo: A Grey-Box Approach to the Functional Testing of Complex Automatic Train Protection Systems. In: Lecture Notes in Computer Science (LNCS) Vol. 3463 (ed. Springer-Verlag Heidelberg): *The Fifth European Dependable Computing Conference*, EDCC-5, Budapest, Hungary, April 20-22, 2005: pp.305-317.

[79] G. Franceschinis, M. Gribaudo, M. Iacono, N. Mazzocca, V. Vittorini: Towards an Object Based Multiformalism Multi-Solution Modeling Approach. In *Proc. of the 2nd Workshop on Modelling of Objects, Components and Agents* (MOCA02), Aarhus, DK, August 2002.

[80] G. Franceschinis, M. Gribaudo, M. Iacono, V. Vittorini, C. Bertoncello: DrawNet++: a flexible framework for building dependability models. In *Proc. of the Int. Conf. on Dependable Systems and Networks*, Washington DC, USA, June 2002.

[81] G. Franceschinis, R. Gaeta, G. Portinaie: Dependability Assessment of an Industrial Programmable Logic Controller via Parametric Fault-Tree and High Level Petri Net. In *Proc. 9th Int. Workshop on Petri Nets and Performance Models*, Aachen, Germany, Sept. 2001, pp. 29-38

[82] G. J. Myers: *The Art of Software Testing.* Wiley, New York (1979).
[83] Ganesh J. Pai, Joanne Bechta Dugan: Automatic Synthesis of Dynamic Fault Trees from UML System Models. In *Proc. 13th International Symposium on Software Reliability Engineering* (ISSRE'02), 2002: p. 243.
[84] Giuliana Franceschinis, Valeria Vittorini, Stefano Marrone, Nicola Mazzocca, SWN Client-Server Composition Operators in the OsMoSys framework. In *IEEE Proceedings of the 10th International Workshop on Petri Nets and Performance Models (PNPM),* 2003, p. 52.
[85] Gregor Gössler and Joseph Sifakis, Composition for Component-Based Modeling. In *proceedings of FMCO'02,* November 5-8, Leiden, the Nederlands.
[86] H. Bohnenkamp, T. Courtney, D. Daly, S. Derisavi, H. Hermanns, J.-P. Katoen, R. Klaren, V. V. Lam, W. H. Sanders. On Integrating the Möbius and Modest Modeling Tools. *In Proc. Int. Conf. on Dependable Systems and Networks,* San Francisco, CA, June 22-25, 2003, p. 671.
[87] H. Pham, H. Wang: Imperfect Maintenance. In *European Journal of Operational Research,* Vol. 94, 1996, pp. 425-438.
[88] Havelund, K., Lowry, M., and Penix, J. 2001. Formal Analysis of a Space-Craft Controller Using SPIN. *IEEE Trans. Softw. Eng.* 27, 8 (Aug. 2001), 749-765.
[89] Hugin Expert home page: http://www.hugin.com/
[90] I. Sommerville: *Software Engineering,* 6th Edition. Addison Wesley (2000).
[91] IBM Rational Rose Real-Time Development Studio Home Page: http://www-306.ibm.com/software/awdtools/suite/technical/
[92] International Electrotechnical Commission: IEC 61508:2000, Parts 1-7, Functional Safety of Electrical/ Electronic/ Programmable Electronic Safety-Related Systems, 2000.
[93] IPL Cantata Web Site: http://www.ipl.com/products/tools/pt400.php

Performance and Dependability Symposium (IPDS'96), IEEE Computer Society Press, 1996, page 60.

[76] G. De Nicola, P. di Tommaso, R. Esposito, F. Flammini, A. Orazzo: A Hybrid Testing Methodology for Railway Control Systems. In: *Lecture Notes in Computer Science (LNCS)* Vol. 3219 (ed. Springer-Verlag Heidelberg): Computer Safety, Reliability, and Security: 23rd International Conference, SAFECOMP 2004, Potsdam, Germany, September 21-24, 2004: pp.116-135.

[77] G. De Nicola, P. di Tommaso, R. Esposito, F. Flammini, P. Marmo, A. Orazzo: ERTMS/ETCS: Working Principles and Validation. In: *Proceedings of the International Conference on Ship Propulsion and Railway Traction Systems,* SPRTS 2005, Bologna, Italy, October 4-6, 2005: pp.59-68.

[78] G. De Nicola, P. di Tommaso, R. Esposito, F. Flammini, P. Marmo, A. Orazzo: A Grey-Box Approach to the Functional Testing of Complex Automatic Train Protection Systems. In: Lecture Notes in Computer Science (LNCS) Vol. 3463 (ed. Springer-Verlag Heidelberg): *The Fifth European Dependable Computing Conference,* EDCC-5, Budapest, Hungary, April 20-22, 2005: pp.305-317.

[79] G. Franceschinis, M. Gribaudo, M. Iacono, N. Mazzocca, V. Vittorini: Towards an Object Based Multiformalism Multi-Solution Modeling Approach. In *Proc. of the 2nd Workshop on Modelling of Objects, Components and Agents* (MOCA02), Aarhus, DK, August 2002.

[80] G. Franceschinis, M. Gribaudo, M. Iacono, V. Vittorini, C. Bertoncello: DrawNet++: a flexible framework for building dependability models. In *Proc. of the Int. Conf. on Dependable Systems and Networks,* Washington DC, USA, June 2002.

[81] G. Franceschinis, R. Gaeta, G. Portinaie: Dependability Assessment of an Industrial Programmable Logic Controller via Parametric Fault-Tree and High Level Petri Net. In *Proc. 9th Int. Workshop on Petri Nets and Performance Models,* Aachen, Germany, Sept. 2001, pp. 29-38

[82] G. J. Myers: *The Art of Software Testing.* Wiley, New York (1979).
[83] Ganesh J. Pai, Joanne Bechta Dugan: Automatic Synthesis of Dynamic Fault Trees from UML System Models. In *Proc. 13th International Symposium on Software Reliability Engineering* (ISSRE'02), 2002: p. 243.
[84] Giuliana Franceschinis, Valeria Vittorini, Stefano Marrone, Nicola Mazzocca, SWN Client-Server Composition Operators in the OsMoSys framework. In *IEEE Proceedings of the 10th International Workshop on Petri Nets and Performance Models (PNPM),* 2003, p. 52.
[85] Gregor Gössler and Joseph Sifakis, Composition for Component-Based Modeling. In *proceedings of FMCO'02,* November 5-8, Leiden, the Nederlands.
[86] H. Bohnenkamp, T. Courtney, D. Daly, S. Derisavi, H. Hermanns, J.-P. Katoen, R. Klaren, V. V. Lam, W. H. Sanders. On Integrating the Möbius and Modest Modeling Tools. *In Proc. Int. Conf. on Dependable Systems and Networks,* San Francisco, CA, June 22-25, 2003, p. 671.
[87] H. Pham, H. Wang: Imperfect Maintenance. In *European Journal of Operational Research,* Vol. 94, 1996, pp. 425-438.
[88] Havelund, K., Lowry, M., and Penix, J. 2001. Formal Analysis of a Space-Craft Controller Using SPIN. *IEEE Trans. Softw. Eng.* 27, 8 (Aug. 2001), 749-765.
[89] Hugin Expert home page: http://www.hugin.com/
[90] I. Sommerville: *Software Engineering,* 6th Edition. Addison Wesley (2000).
[91] IBM Rational Rose Real-Time Development Studio Home Page: http://www-306.ibm.com/software/awdtools/suite/technical/
[92] International Electrotechnical Commission: IEC 61508:2000, Parts 1-7, Functional Safety of Electrical/ Electronic/ Programmable Electronic Safety-Related Systems, 2000.
[93] IPL Cantata Web Site: http://www.ipl.com/products/tools/pt400.php

Performance and Dependability Symposium (IPDS'96), IEEE Computer Society Press, 1996, page 60.

[76] G. De Nicola, P. di Tommaso, R. Esposito, F. Flammini, A. Orazzo: A Hybrid Testing Methodology for Railway Control Systems. In: *Lecture Notes in Computer Science (LNCS)* Vol. 3219 (ed. Springer-Verlag Heidelberg): Computer Safety, Reliability, and Security: 23rd International Conference, SAFECOMP 2004, Potsdam, Germany, September 21-24, 2004: pp.116-135.

[77] G. De Nicola, P. di Tommaso, R. Esposito, F. Flammini, P. Marmo, A. Orazzo: ERTMS/ETCS: Working Principles and Validation. In: *Proceedings of the International Conference on Ship Propulsion and Railway Traction Systems,* SPRTS 2005, Bologna, Italy, October 4-6, 2005: pp.59-68.

[78] G. De Nicola, P. di Tommaso, R. Esposito, F. Flammini, P. Marmo, A. Orazzo: A Grey-Box Approach to the Functional Testing of Complex Automatic Train Protection Systems. In: Lecture Notes in Computer Science (LNCS) Vol. 3463 (ed. Springer-Verlag Heidelberg): *The Fifth European Dependable Computing Conference,* EDCC-5, Budapest, Hungary, April 20-22, 2005: pp.305-317.

[79] G. Franceschinis, M. Gribaudo, M. Iacono, N. Mazzocca, V. Vittorini: Towards an Object Based Multiformalism Multi-Solution Modeling Approach. In *Proc. of the 2nd Workshop on Modelling of Objects, Components and Agents* (MOCA02), Aarhus, DK, August 2002.

[80] G. Franceschinis, M. Gribaudo, M. Iacono, V. Vittorini, C. Bertoncello: DrawNet++: a flexible framework for building dependability models. In *Proc. of the Int. Conf. on Dependable Systems and Networks,* Washington DC, USA, June 2002.

[81] G. Franceschinis, R. Gaeta, G. Portinaie: Dependability Assessment of an Industrial Programmable Logic Controller via Parametric Fault-Tree and High Level Petri Net. In *Proc. 9th Int. Workshop on Petri Nets and Performance Models,* Aachen, Germany, Sept. 2001, pp. 29-38

[82] G. J. Myers: *The Art of Software Testing.* Wiley, New York (1979).
[83] Ganesh J. Pai, Joanne Bechta Dugan: Automatic Synthesis of Dynamic Fault Trees from UML System Models. In *Proc. 13th International Symposium on Software Reliability Engineering* (ISSRE'02), 2002: p. 243.
[84] Giuliana Franceschinis, Valeria Vittorini, Stefano Marrone, Nicola Mazzocca, SWN Client-Server Composition Operators in the OsMoSys framework. In *IEEE Proceedings of the 10th International Workshop on Petri Nets and Performance Models (PNPM),* 2003, p. 52.
[85] Gregor Gössler and Joseph Sifakis, Composition for Component-Based Modeling. In *proceedings of FMCO'02,* November 5-8, Leiden, the Nederlands.
[86] H. Bohnenkamp, T. Courtney, D. Daly, S. Derisavi, H. Hermanns, J.-P. Katoen, R. Klaren, V. V. Lam, W. H. Sanders. On Integrating the Möbius and Modest Modeling Tools. *In Proc. Int. Conf. on Dependable Systems and Networks,* San Francisco, CA, June 22-25, 2003, p. 671.
[87] H. Pham, H. Wang: Imperfect Maintenance. In *European Journal of Operational Research,* Vol. 94, 1996, pp. 425-438.
[88] Havelund, K., Lowry, M., and Penix, J. 2001. Formal Analysis of a Space-Craft Controller Using SPIN. *IEEE Trans. Softw. Eng.* 27, 8 (Aug. 2001), 749-765.
[89] Hugin Expert home page: http://www.hugin.com/
[90] I. Sommerville: *Software Engineering,* 6th Edition. Addison Wesley (2000).
[91] IBM Rational Rose Real-Time Development Studio Home Page: http://www-306.ibm.com/software/awdtools/suite/technical/
[92] International Electrotechnical Commission: IEC 61508:2000, Parts 1-7, Functional Safety of Electrical/ Electronic/ Programmable Electronic Safety-Related Systems, 2000.
[93] IPL Cantata Web Site: http://www.ipl.com/products/tools/pt400.php

[94] J. Arlat, K. Kanoun and J.C. Laprie, Dependability Modeling and Evalutation of Software Fault-Tolerant Systems, *IEEE TC*, Vol. C-39, pp. 504-512, 1990.

[95] J. B. Dugan, S. J. Bavoso, M. A. Boyd, "Dynamic Fault-Tree Models for Fault Tolerant Computer Systems", *IEEE Transactions on Reliability,* vol. 41, 1992, pp. 363-377.

[96] J. B. Dugan, S. J. Bavuso, M. A. Boyd: Dynamic Fault-Tree Models for Fault-Tolerant Computer Systems. In *IEEE Transactions on Reliability,* vol. 41, 1992, pp. 363-377.

[97] J. M. Wing, A Specifier's Introduction to Formal Methods. In *IEEE Computer*, Vol. 23, No. 9, September 1990, pp. 8-24.

[98] J. Wegener, K. Grimm, M. Grochtmann: *Systematic Testing of Real-Time Systems.* Conference Papers of EuroSTAR '96, Amsterdam (1996).

[99] J.-C. Laprie (Ed.), Dependability: Basic Concepts and Terminology, *Dependable Computing and Fault-Tolerance,* 5, 265p., Springer-Verlag, Vienna, Austria, 1992.

[100] J.-C. Laprie, Dependable Computing: Concepts, Limits, Challenges, in Special Issue, 25th IEEE Int. *Symp. on Fault-Tolerant Computing* (FTCS- 25), (Pasadena, CA, USA), pp.42-54, IEEE Computer Society Press, 1995.

[101] J.-C. Laprie, J. Arlat, J.-P. Blanquart, A. Costes, Y. Crouzet, Y. Deswarte, J.-C. Fabre, H. Guillermain, M. Kaâniche, K. Kanoun, C. Mazet, D. Powell, C. Rabéjac and P. Thévenod, *Dependability Handbook,* Report N°98346, LAAS-CNRS, 1998 (draft).

[102] J.F. Meyer. Performability: a retrospective and some pointers to the future. *Performance Evaluation,* vol 14(3&4): 139–156. Elsevier. 1992.

[103] J.J. McCall: Maintenance Policies for Stochastically Failing Equipment: A Survey. In *Management Science,* Vol. 11, No. 5, 1965, pp. 493-524.

[104] Jansen L., Meyer zu Horste M., Schneider E. Technical issues in modelling the European Train Control System (ETCS) using coloured Petri nets and the Design/CPN tools. In *Proc. 1st CPN*

Workshop, DAIMI PB 532, pages 103--115. Aarhus University, 1998.
[105] Jean Arlat, Nobuyasu Kanekawa, Arturo M. Amendola, Jean-Luis Dufour, Yuji Hirao, Joseph A. Profeta III: "Dependability of Railway Control Systems". *FTCS* 1996: 150-155.
[106] K. A. Delic, F. Mazzanti, L. Strigini, Formalizing Engineering Judgement on Software Dependability via Belief Networks, *6th IFIP Working Conference on Dependable Computing for Critical Applications,* 1997.
[107] K. Grimm: Systematic Testing of Software-Based Systems. In *Proceedings of the 2nd Annual ENCRESS Conference*, Paris (1996)
[108] K. Jensen, *Coloured Petri Nets: Basic Concepts, Analysis Methods and Practical* Use, vol. I, Springer Verlag, 1992.
[109] Kevin Murphy: Software Packages for Bayesian Networks, http://www.cs.ubc.ca/~murphyk/Bayes/bnsoft.html
[110] L. Portinale, A. Bobbio, S. Montani, "From AI to Dependability: Using Bayesian Networks for Reliability Modeling and Analysis". In *Fourth International Conference on Mathematical Methods in Reliability* (MMR2004), June 2004.
[111] Lara, J. d. and Vangheluwe, H. 2002. AToM3: A Tool for Multi-formalism and Meta-modelling. In *Proceedings of the 5th international Conference on Fundamental Approaches To Software Engineering* (April 08 - 12, 2002). R. Kutsche and H. Weber, Eds. Lecture Notes In Computer Science, vol. 2306. Springer-Verlag, London, 174-188.
[112] M. Ajmone Marsan, G. Balbo, G. Conte, S. Donatelli, G. Franceschinis: *Modelling with Generalized Stochastic Petri Nets.* J. Wiley and Sons ed., 1995.
[113] M. Gribaudo, M. Iacono, N. Mazzocca, V. Vittorini: The OsMoSys/DrawNET Xe! Languages System: A Novel Infrastructure for Multiformalism Object-Oriented Modelling. In *Proc. 15th European Simulation Symposium and Exhibition,* Delft, The Netherlands, October 2003.

[114] M. Grochtmann, K. Grimm: Classification-Trees for Partition Testing. *Journal of Software Testing,* Verification and Reliability, Vol. 3, No.2, (1993) 63-82.
[115] M. Hsueh, T. Tsai, R. Iyer: Fault injection techniques and tools. In *IEEE Computer,* vol. 30, no. 4, April 1997, pp. 75-82.
[116] M. Iacono: "A multiformalism methodology for heterogeneous computer systems analysis". *PhD thesis,* 2002.
[117] M.J.M. Houtermans: *Safety Management Tools and Techniques,* http://www.isa.org/~safety/safety_management_tools.htm
[118] Martha A. Centeno: An introduction to simulation modeling. In *Proceedings of the 28th conference on Winter Simulation Conference,* Coronado, California, United States, 1996, pp. 15 - 22
[119] Martin Fowler et al.: Refactoring: *Improving the Design of Existing Code,* Addison-Wesley Professional, 1st edition (1999).
[120] Martin Fowler: UML Distilled: *A Brief Guide to the Standard Object Modeling Language* (Object Technology S.), Addison Wesley, 2004.
[121] Massoud Amin, "Infrastructure Security: Reliability and Dependability of Critical Systems". In *IEEE Security & Privacy,* vol. 3, no. 3, 2005, pp. 15-17.
[122] Merz, S.: *Model checking: a tutorial overview.* In Modeling and Verification of Parallel Processes, F. Cassez, C. Jard, B. Rozoy, and M. D. Ryan, Eds. Lecture Notes In *Computer Science,* vol. 2067. Springer-Verlag New York, New York, NY, 2001, pp: 3-38.
[123] Model-Based Testing Home Page: www.model-based-testing.org/
[124] Norman B. Fuqua, *Reliability Engineering for Electronic Design,* Marcel Dekker Inc, 1987.
[125] Norsys Netica home page: http://www.norsys.com/netica.html
[126] Ntafos, Simeon,"A Comparison of Some Structural Testing Strategies", *IEEE Trans. Software Eng.,* Vol.14, No.6, June 1988, pp.868-874.
[127] OMG Unified Modeling Language Home Page: http://www.uml.org/
[128] P. L. Clemens, *Fault Tree Analysis,* 4th Edition.
[129] Peter Ball: Introduction to Discrete Event Simulation. In *Proceedings of the 2nd DYCOMANS workshop on "Management*

and Control : Tools in Action" in the Algarve, Portugal. 15th - 17th May 1996, pp. 367-376.

[130] Peter Biechele, Stefan Leue: *Explicit State Model Checking in the Development Process for Interlocking Software Systems.* Extended Abstract, DSN 2003, url = citeseer.ist.psu.edu/biechele03 explicit.html

[131] R. A. Weaver, T. P. Kelly: *The Goal Structuring Notation - A Safety Argument Notation.* In *Proc. of Dependable Systems and Networks 2004 Workshop on Assurance Cases,* July 2004.

[132] R. Dekker: Applications of Maintenance Optimization Models: A Review and Analysis. In *Reliability Engineering and System Safety,* Vol. 51, No. 3, 1996, pp. 229-240.

[133] R. Manian, D. W. Coppit, K. J. Sullivan, J. B. Dugan: Bridging the Gap Between Systems and Dynamic Fault Tree Models. In *Proceedings Annual Reliability and Maintainability Symposium,* 1999, pp. 105-111.

[134] R.A. Sahner and K.S. Trivedi and A. Puliafito. *Performance and Reliability Analysis of Computer Systems: An Example-based Approach Using the SHARPE Software Package,* Kluwer Academic Publishers, 1996.

[135] R.E. Barlow and F. Proschan, *Mathematical Theory of Reliability,* John Wiley & Sons, Inc., New York, 1965.

[136] Roper, Marc, *"Software Testing",* London, McGraw-Hill Book Company, 1994.

[137] RTCA SC-167, EUROCAE WG-12: DO-178B / ED-12B - Software Considerations in Airborne Systems and Equipment Certification (1992).

[138] S. Chiaradonna, A. Bondavalli and L. Strigini, On Performability Modeling and Evaluation of Software Fault Tolerance Structures, In *Proc. of EDCC-1,* LNCS 852, Springer Verlag, 1994, pp. 97-114.

[139] S. Gnesi, G. Lendini, C. Abbaneo, D. Latella, A. Amendola and P. Marmo: An Automatic SPIN Validation of a Safety Critical Railway Control System. In *Proc. of Int. Conf. on Dependable Systems and Networks,* (2000): pp. 119-124.

[140] S. Kent. Model Driven Engineering. In *Integrated Formal Methods.* Vol. 2335 of LNCS. Springer-Verlag, 2002.

[141] S. Montani, L. Portinale, A. Bobbio: "Dynamic Bayesian Networks for Modeling Advanced Fault Tree Features in Dependability Analysis". In *Proc. European Safety and Reliability Conference* (ESREL 2005), Tri City, Poland, 2005: pp. 1415-1422.

[142] S. Osaki, T. Nakagawa. Bibliography for Reliability and Availability of Stochastic Systems. In *IEEE Transactions on Reliability*, vol. 25, 1976, pp. 284-287.

[143] S.M. Ross, Introduction to Probability Models, Seventh Edition, *Harcourt Academic Press,* San Diego, 1989.

[144] Slavisa Markovic: *Composition of UML Described Refactoring Rules.* In *OCL and Model Driven Engineering,* UML 2004 Conference Workshop, October 12, 2004, Lisbon, Portugal, pp. 45-59.

[145] Stefania Gnesi, Diego Latella, Gabriele Lenzini, C. Abbaneo, Arturo M. Amendola, P. Marmo: "A Formal Specification and Validation of a Critical System in Presence of Byzantine Errors". *TACAS* 2000: 535-549.

[146] T. L. Graves, M. J. Harrold, J. M. Kim, A. Porter, G. Rothermel: An Empirical Study of Regression Test Selection Techniques. In *Proceedings of the 20th International Conference on Software Engineering* (1998) 188-197.

[147] T. Ostrand, M. Balcer: The Category-Partition Method for Specifying and Generating Functional Tests. *Communications of the ACM,* 31 (6), (1988) 676-686.

[148] *Telelogic Tau Logicscope* v5.1: Basic Concept. (2001).

[149] TÜV Rheinland Group Home Page: https://www.tuv.com

[150] U. S. Nuclear Regulatory Commission, *Fault Tree Handbook,* NUREG-0492, 1981.

[151] UIC, *ERTMS/ETCS class1 System Requirements Specification,* Ref. SUBSET-026, issue 2.2.2, 2002.

[152] UIC, *ERTMS/ETCS class1 System Requirements Specification,* Ref. SUBSET-076, issue 2.2.3, 2005.

[153] UNISIG, *ERTMS/ETCS RAMS Requirements Specification*, Ref. 96s1266.
[154] V. Vittorini, M. Iacono, N. Mazzocca, G. Franceschinis, The OsMoSys approach to multiformalism modeling of systems. In *Journal of Software and Systems Modeling,* Volume 3, Issue 1, March 2004: pp. 68-81.
[155] W. G. Bouricius, W. C. Carter, P. R. Schneider, *Reliability Modeling Techniques for Self Repairing Computer Systems,* IBM Watson Research Center, New York.
[156] W. S. Heath: *Real-Time Software Techniques.* Van Nostrand Reinhold, New York (1991).
[157] W.F. Rice, C.R. Cassady, J.A. Nachlas: Optimal Maintenance Plans under Limited Maintenance Time. In *Industrial Engineering Research '98 Conference Proceedings,* 1998.
[158] Wiboonsak Watthayu et al.: "A Bayesian network based framework for multi-criteria decision making". In *Proceedings of the 17th International Conference on Multiple Criteria Decision Analysis,* August 2004.
[159] Y.S. Barlow, M.L. Smith: Optimal Maintenance Models for Systems Subject to Failure - A Review. In *Naval Research Logistics Quarterly,* Vol. 28, 1981, pp. 47-74.
[160] YELLOW BOOK Home Page: http://www.yellowbook-rail.org.uk

INDEX

A

Abstraction, 11
actuators, 12, 32
ADC, 13
aerospace, 2, 8, 11, 24
artificial intelligence, 47
assessment, 2, 23
assignment, 25
ATP, 51
automation, 16, 59
availability, 8, 11, 12, 20, 30, 34, 35, 39, 43, 44, 45

B

basic needs, 14
behavior, 1, 10, 17, 41
behavioral aspects, 46
blocks, 10, 15, 45

C

certification, 22, 25
City, 56, 69
classes, 1, 8, 10, 35, 39, 41, 42
classification, 7, 17, 21, 27, 35
coding, 9, 17, 36
coherence, 19

communication, 3, 13, 15, 30, 36, 60
complex behaviors, 44
complexity, 3, 11, 43, 47
components, 12, 15, 18, 20, 22, 26, 33, 34, 45, 49
computer research, 3
computer systems, 5, 6, 7, 43, 67
computing, 7, 28
configuration, 11, 12, 16, 18, 20, 21, 31, 32
confinement, 13
consensus, 28
construction, 36
control, 2, 3, 5, 6, 7, 8, 9, 11, 12, 13, 17, 22, 29, 32, 44, 60
CPU, 10, 11, 12, 18, 51

D

database, 32
decision making, 2, 70
decomposition, 58
definition, 7, 14, 15, 21, 25, 42
density, 28, 29
Department of Defense, 24
detection, 18, 22
deviation, 26
DFT, 44, 52
distribution, 16, 32

diversity, 18, 23
division, 2, 30, 37

E

electromagnetic, 20
electromagnetic fields, 20
electronic systems, 30, 60
encapsulation, 26
environment, 1, 13, 21, 22, 28, 40, 41
environmental conditions, 8, 20
environmental impact, 2
ergonomics, 14
error detection, 12
evolution, 41
execution, 21, 28, 41, 42
external environment, 13

F

failure, 1, 2, 5, 6, 7, 12, 22, 23, 24, 25, 26, 32, 33, 43
fault tolerance, 8, 10, 18
feedback, 28, 41
flexibility, 40, 49
formal language, 16, 36, 43
France, iii, 62

G

Germany, 61, 62, 63
graph, 41
guidelines, 22, 23, 24, 25, 32

H

HDL, 20, 40, 52
Hungary, 56, 57, 63
hypothesis, 32

I

implementation, 8, 15, 16, 18, 27, 36, 37
indicators, 28
indices, 25, 27, 32, 33
infant mortality, 32
integration, 15, 27, 31, 32, 37, 40, 41, 42
integrity, 14, 23, 25, 27

interaction, 9, 26
interactions, 15
interface, 13, 15, 41, 42
international standards, 16
interoperability, 15
intuition, 11, 34
isolation, 13
Italy, 57, 59, 60, 63

J

Java, 26, 48, 59

L

language, 15, 16, 17, 26, 35, 36, 46
life cycle, 15
limitation, 44, 46
logistics, 21, 34

M

maintenance, 12, 17, 21, 34, 37, 44, 60
management, 12, 15, 17, 21, 22, 23, 24, 25, 26, 28, 35, 36, 67
market, 12
measures, 20, 28
memory, 10, 12, 20, 26
military, 18
model, 13, 14, 18, 21, 25, 27, 28, 32, 35, 36, 39, 40, 42, 43, 44, 45, 46, 47, 48, 49, 67
modeling, 43, 45, 46, 47, 48, 49, 67, 70
models, 9, 20, 27, 30, 40, 47, 48, 59, 62, 63
modules, 3, 13, 16, 20, 26, 27, 37, 41
motivation, 43

N

NATO, 55
Netherlands, 66
network, 3, 47, 70
neural network, 23
neural networks, 23

Index

O

operating system, 10, 13, 19
Operators, 64
optimization, 47
order, 8, 10, 12, 13, 14, 16, 17, 19, 20, 21, 26, 34, 36, 39, 40, 41, 42

P

parameters, 21, 27, 36, 37
Poland, 59, 69
Portugal, 62, 68, 69
power, 12, 19, 34, 36, 43, 44, 45, 46, 49
predictability, 7, 17
prediction, 27, 35, 36, 37, 38, 39
probability, 2, 9, 10, 11, 20, 47
programming, 10, 17, 47
programming languages, 47
protocol, 13, 15, 36
prototype, 39, 40

Q

quality standards, 17

R

range, 18, 26
recovery, 10, 12, 18
redundancy, 10, 12, 18, 36
reliability, 10, 12, 15, 19, 20, 21, 24, 30, 33, 36, 37, 39, 43, 44, 45, 47, 49
repair, 12, 19, 21, 33, 34, 44, 46, 56
Requirements, 14, 15, 16, 25, 52, 54, 58, 69, 70
robustness, 12, 20, 42

S

safety, 2, 3, 5, 8, 9, 10, 11, 14, 15, 19, 22, 23, 24, 25, 26, 27, 29, 30, 31, 32, 35, 39, 42, 43, 67

sensors, 12, 32
signalling, 29, 30, 60
signals, 11
simulation, 21, 37, 39, 40, 41, 48, 67
software, 5, 8, 10, 11, 13, 14, 15, 16, 17, 18, 20, 22, 23, 26, 27, 28, 31, 36, 37, 39, 40, 41, 43, 46, 47, 48, 56, 64
space, 11, 46
stability, 48
standards, 15, 19, 22, 23, 24, 25, 26, 29
STM, 54
strategies, 12, 34, 44, 46
survivability, 7, 10, 14, 20

T

threats, 9, 21
time constraints, 7
timing, 7, 11, 17
transitions, 46
translation, 45, 47
transmission, 30, 60
transport, 13
transportation, 2, 12
trustworthiness, 2, 22

U

United States, 67

V

validation, 19, 22, 24, 27, 31, 34, 35, 36
variables, 18, 41, 47
voting, 10, 18, 19

W

wear, 12, 32
workload, 24, 25

X

XML, 54